RADIATION PROTECTION AND SAFETY OF RADIATION SOURCES: INTERNATIONAL BASIC SAFETY STANDARDS

INTERIM EDITION

The following States are Members of the International Atomic Energy Agency:

AFGHANISTAN
ALBANIA
ALGERIA
ANGOLA
ARGENTINA
ARMENIA
AUSTRALIA
AUSTRIA
AZERBAIJAN
BAHRAIN
BANGLADESH
BELARUS
BELGIUM
BELIZE
BENIN
BOLIVIA
BOSNIA AND HERZEGOVINA
BOTSWANA
BRAZIL
BULGARIA
BURKINA FASO
BURUNDI
CAMBODIA
CAMEROON
CANADA
CENTRAL AFRICAN REPUBLIC
CHAD
CHILE
CHINA
COLOMBIA
CONGO
COSTA RICA
CÔTE D'IVOIRE
CROATIA
CUBA
CYPRUS
CZECH REPUBLIC
DEMOCRATIC REPUBLIC OF THE CONGO
DENMARK
DOMINICAN REPUBLIC
ECUADOR
EGYPT
EL SALVADOR
ERITREA
ESTONIA
ETHIOPIA
FINLAND
FRANCE
GABON
GEORGIA
GERMANY
GHANA
GREECE
GUATEMALA
HAITI
HOLY SEE
HONDURAS
HUNGARY
ICELAND
INDIA
INDONESIA
IRAN, ISLAMIC REPUBLIC OF
IRAQ
IRELAND
ISRAEL
ITALY
JAMAICA
JAPAN
JORDAN
KAZAKHSTAN
KENYA
KOREA, REPUBLIC OF
KUWAIT
KYRGYZSTAN
LATVIA
LEBANON
LESOTHO
LIBERIA
LIBYA
LIECHTENSTEIN
LITHUANIA
LUXEMBOURG
MADAGASCAR
MALAWI
MALAYSIA
MALI
MALTA
MARSHALL ISLANDS
MAURITANIA
MAURITIUS
MEXICO
MONACO
MONGOLIA
MONTENEGRO
MOROCCO
MOZAMBIQUE
MYANMAR
NAMIBIA
NEPAL
NETHERLANDS
NEW ZEALAND
NICARAGUA
NIGER
NIGERIA
NORWAY
OMAN
PAKISTAN
PALAU
PANAMA
PARAGUAY
PERU
PHILIPPINES
POLAND
PORTUGAL
QATAR
REPUBLIC OF MOLDOVA
ROMANIA
RUSSIAN FEDERATION
SAUDI ARABIA
SENEGAL
SERBIA
SEYCHELLES
SIERRA LEONE
SINGAPORE
SLOVAKIA
SLOVENIA
SOUTH AFRICA
SPAIN
SRI LANKA
SUDAN
SWEDEN
SWITZERLAND
SYRIAN ARAB REPUBLIC
TAJIKISTAN
THAILAND
THE FORMER YUGOSLAV REPUBLIC OF MACEDONIA
TUNISIA
TURKEY
UGANDA
UKRAINE
UNITED ARAB EMIRATES
UNITED KINGDOM OF GREAT BRITAIN AND NORTHERN IRELAND
UNITED REPUBLIC OF TANZANIA
UNITED STATES OF AMERICA
URUGUAY
UZBEKISTAN
VENEZUELA
VIETNAM
YEMEN
ZAMBIA
ZIMBABWE

The Agency's Statute was approved on 23 October 1956 by the Conference on the Statute of the IAEA held at United Nations Headquarters, New York; it entered into force on 29 July 1957. The Headquarters of the Agency are situated in Vienna. Its principal objective is "to accelerate and enlarge the contribution of atomic energy to peace, health and prosperity throughout the world".

IAEA SAFETY STANDARDS SERIES No. GSR Part 3 (Interim)

RADIATION PROTECTION AND SAFETY OF RADIATION SOURCES: INTERNATIONAL BASIC SAFETY STANDARDS

INTERIM EDITION

GENERAL SAFETY REQUIREMENTS

This publication includes a CD-ROM containing the IAEA Safety Glossary: 2007 Edition (2007), the Fundamental Safety Principles (2006) and Radiation Protection and Safety of Radiation Sources: International Basic Safety Standards: Interim Edition (2011), each in Arabic, Chinese, English, French, Russian and Spanish versions.

INTERNATIONAL ATOMIC ENERGY AGENCY
VIENNA, 2011

Marketing and Sales Unit, Publishing Section
International Atomic Energy Agency
Vienna International Centre
PO Box 100
1400 Vienna, Austria
fax: +43 1 2600 29302
tel.: +43 1 2600 22417
email: sales.publications@iaea.org
http://www.iaea.org/books

Printed by the IAEA in Austria
November 2011
STI/PUB/1531

IAEA Library Cataloguing in Publication Data

Radiation protection and safety of radiation sources : international basic safety standards : general safety requirements. – Interim edition. – Vienna : International Atomic Energy Agency, 2011.
p. ; 29 cm. – (IAEA safety standards series, ISSN 1020-525X ; no. GRS Part 3 (Interim))
STI/PUB/1531
ISBN 978–92–0–120910–8
Includes bibliographical references.

1. Nuclear facilities — Safety measures. 2. Radiation — Safety measures. 3. Radiation protection. 4. Safety standards. I. International Atomic Energy Agency. II. Series.

IAEAL 11–00705

FOREWORD

by Yukiya Amano
Director General

The IAEA's Statute authorizes the Agency to "establish or adopt… standards of safety for protection of health and minimization of danger to life and property" — standards that the IAEA must use in its own operations, and which States can apply by means of their regulatory provisions for nuclear and radiation safety. The IAEA does this in consultation with the competent organs of the United Nations and with the specialized agencies concerned. A comprehensive set of high quality standards under regular review is a key element of a stable and sustainable global safety regime, as is the IAEA's assistance in their application.

The IAEA commenced its safety standards programme in 1958. The emphasis placed on quality, fitness for purpose and continuous improvement has led to the widespread use of the IAEA standards throughout the world. The Safety Standards Series now includes unified Fundamental Safety Principles, which represent an international consensus on what must constitute a high level of protection and safety. With the strong support of the Commission on Safety Standards, the IAEA is working to promote the global acceptance and use of its standards.

Standards are only effective if they are properly applied in practice. The IAEA's safety services encompass design, siting and engineering safety, operational safety, radiation safety, safe transport of radioactive material and safe management of radioactive waste, as well as governmental organization, regulatory matters and safety culture in organizations. These safety services assist Member States in the application of the standards and enable valuable experience and insights to be shared.

Regulating safety is a national responsibility, and many States have decided to adopt the IAEA's standards for use in their national regulations. For parties to the various international safety conventions, IAEA standards provide a consistent, reliable means of ensuring the effective fulfilment of obligations under the conventions. The standards are also applied by regulatory bodies and operators around the world to enhance safety in nuclear power generation and in nuclear applications in medicine, industry, agriculture and research.

Safety is not an end in itself but a prerequisite for the purpose of the protection of people in all States and of the environment — now and in the future. The risks associated with ionizing radiation must be assessed and controlled without unduly limiting the contribution of nuclear energy to equitable and sustainable development. Governments, regulatory bodies and operators everywhere must ensure that nuclear material and radiation sources are used beneficially, safely and ethically. The IAEA safety standards are designed to facilitate this, and I encourage all Member States to make use of them.

NOTE BY THE SECRETARIAT

The IAEA safety standards reflect an international consensus on what constitutes a high level of safety for protecting people and the environment from harmful effects of ionizing radiation. The process of developing, reviewing and establishing the IAEA standards involves the IAEA Secretariat and all Member States, many of which are represented on the four IAEA safety standards committees and the IAEA Commission on Safety Standards.

The IAEA standards, as a key element of the global safety regime, are kept under regular review by the Secretariat, the safety standards committees and the Commission on Safety Standards. The Secretariat gathers information on experience in the application of the IAEA standards and information gained from the follow-up of events for the purpose of ensuring that the standards continue to meet users' needs. The present publication reflects feedback and experience accumulated until 2010 and it has been subject to the rigorous review process for standards.

Lessons that may be learned from studying the accident at the Fukushima Daiichi nuclear power plant in Japan following the disastrous earthquake and tsunami of 11 March 2011 will be reflected in this IAEA Safety Requirements publication as revised and issued in the future.

PREFACE TO THE INTERIM EDITION

The Board of Governors of the IAEA, at its meeting on 12 September 2011, established as an IAEA safety standard — in accordance with Article III.A.6 of the Statute — this Safety Requirements publication, and authorized the Director General to promulgate these Safety Requirements and to issue them as a Safety Requirements publication in the IAEA Safety Standards Series.

The Board of Governors of the IAEA first approved basic safety standards in June 1962, which were published by the IAEA as Safety Series No. 9. A revised version was published in 1967. A third revision was published by the IAEA as the 1982 Edition of Safety Series No. 9; this edition was jointly sponsored by the IAEA, International Labour Organization (ILO), OECD Nuclear Energy Agency (OECD/NEA) and World Health Organization (WHO). The next version was the International Basic Safety Standards for Protection against Ionizing Radiation and for the Safety of Radiation Sources (the BSS), published by the IAEA in February 1996 as Safety Series No. 115 and jointly sponsored by the Food and Agriculture Organization of the United Nations (FAO), IAEA, ILO, OECD/NEA, Pan American Health Organization (PAHO) and WHO.

In September 2005, the General Conference of the IAEA, by resolution GC(49)/RES/9, requested the IAEA Secretariat to undertake a review of the BSS. The IAEA invited representatives of United Nations and other intergovernmental organizations to participate in the review and revision of the BSS through the establishment of a BSS Secretariat made up of designated representatives of the potential sponsoring organizations: the European Commission (EC/Euratom), FAO, ILO, OECD/NEA, PAHO, United Nations Environment Programme (UNEP) and WHO. The BSS Secretariat was supported by the IAEA Secretariat.

In September 2006, the IAEA General Conference, by way of resolution GC(50)/RES/10, noted the review of the BSS carried out in response to para. 10 of resolution GC(49)/RES/9, and noted that the revision was to be coordinated by the BSS Secretariat.

The revision of the BSS commenced in early 2007 with a series of drafting meetings on the basis of thematic areas, hosted by the IAEA, ILO, WHO, OECD/NEA and PAHO. The draft text developed at these meetings provided the basis for discussions at a technical meeting held in July 2007, with the participation of representatives of the potential sponsoring organizations.

The draft text takes account of the findings of the United Nations Scientific Committee on the Effects of Atomic Radiation (UNSCEAR) and the recommendations of the International Commission on Radiological Protection (ICRP).

The technical meeting recommended that the revised BSS be structured on the basis of 'planned exposure situations', 'emergency exposure situations' and 'existing exposure situations', with each main section of the text having the same layout covering occupational exposure, public exposure and (for planned exposure situations only) medical exposure, following the ICRP recommendations. There was also to be a main section dealing with generic requirements that are applicable in all exposure situations. The technical meeting also recommended that the revised BSS cover protection of the environment, to be consistent with the Fundamental Safety Principles, published in 2006 as IAEA Safety Standards Series No. SF-1.

Further drafting and review meetings with the potential sponsoring organizations were held from late 2007 to 2009. The IAEA Safety Standards Committees and groups of experts from potential sponsoring organizations provided feedback on drafts of the revised BSS in 2008 and 2009. A further technical meeting with the participation of representatives of the potential sponsoring organizations was held in December 2009 to discuss a statement on radon that was issued by the ICRP in November 2009 and to assess its implication for the revised BSS. The technical meeting also made recommendations on the text of the revised BSS in relation to exposure to radon indoors and to occupational exposure to radon.

In January 2010, the draft text was submitted to IAEA Member States for comment. On the basis of the comments received, a revised draft text was prepared. The revised draft text was subsequently approved by the Nuclear Safety Standards Committee and the Transport Safety Standards Committee in November 2010, and by the Radiation Safety Standards Committee and the Waste Safety Standards Committee in December 2010, and was endorsed by the Commission on Safety Standards (CSS) in May 2011.

On 21 April 2011, the ICRP issued a statement on tissue reactions which recommended a reduction in the dose limit (in terms of equivalent dose) for the lens of the eye. The CSS concluded in May 2011 that it intends to incorporate the revised dose limit to the lens of the eye into Schedule III of the revised BSS, after consultation of

Member States. Member States were invited to provide comment on the revised Schedule III by 7 July 2011. Following a recommendation from the outgoing and incoming Chairs of the Radiation Safety Standards Committee, the Chair of the CSS approved the revised Schedule III on 12 July 2011.

The objective of this Safety Requirements publication is to establish, on the basis of the safety objective and principles established in the Fundamental Safety Principles, requirements for the protection of people and the environment from harmful effects of ionizing radiation and for the safety of radiation sources.

This publication is intended for use by governmental authorities including regulatory bodies responsible for licensing facilities and activities; by organizations operating nuclear facilities, some mining and raw material processing facilities such as uranium mines, radioactive waste management facilities, and any other facilities producing or using radiation sources for industrial, research or medical purposes; by organizations transporting radioactive material; by organizations decommissioning facilities; and by staff and technical and scientific support organizations supporting such organizations and authorities.

The references included in this publication are to editions that are current as of the time of publication of these Standards. Editions that supersede these references may be adopted under national legislation. In the event that the publications referenced here are superseded, please refer to the most recent editions. See also: *http://www-ns.iaea.org/standards/*.

This publication also includes a list of definitions which apply for the purposes of these Standards. The list of definitions includes: definitions of new terms that are not included in the IAEA Safety Glossary: Terminology Used in Nuclear Safety and Radiation Protection (2007 Edition); revised definitions of terms that are defined in the IAEA Safety Glossary; and existing definitions from the IAEA Safety Glossary, included here for ease of reference. The revised definitions included here of terms with definitions in the IAEA Safety Glossary will be included in the next revision of the IAEA Safety Glossary and will thus supersede those existing definitions. Further relevant definitions are provided in the IAEA Safety Glossary. See also: *http://www-ns.iaea.org/standards/safety-glossary.*

Also included is a CD-ROM containing the IAEA Safety Glossary: 2007 Edition (2007), the Fundamental Safety Principles (2006) and the present Safety Requirements publication, Radiation Protection and Safety of Radiation Sources: International Basic Safety Standards: Interim Edition (2011) (including tables in Schedule III that are not included in the printed book), each in Arabic, Chinese, English, French, Russian and Spanish versions.

The revised BSS are hereby published as General Safety Requirements Part 3 (Interim) in the IAEA Safety Standards Series in an Interim Edition. This Interim Edition has been submitted to the other potential sponsoring organizations for their approval. Following their decisions on its approval, it will be issued as a jointly sponsored standard.

This Interim Edition is reproduced directly from Board of Governors document GOV/2011/42 and has not been edited or typeset by the IAEA. Please notify any apparent errors, omissions or mistranslations by email to *Rad.prot.unit@iaea.org* and *Safety.Standards@iaea.org,* or via the IAEA safety standards web site, *http://www-ns.iaea.org/standards/*, so that any necessary corrections can be made in the final edition.

The IAEA wishes to express its great appreciation to all those who assisted in the drafting, review, revision and translation of this text and in the process of reaching a consensus.

THE IAEA SAFETY STANDARDS

BACKGROUND

Radioactivity is a natural phenomenon and natural sources of radiation are features of the environment. Radiation and radioactive substances have many beneficial applications, ranging from power generation to uses in medicine, industry and agriculture. The radiation risks to workers and the public and to the environment that may arise from these applications have to be assessed and, if necessary, controlled.

Activities such as the medical uses of radiation, the operation of nuclear installations, the production, transport and use of radioactive material, and the management of radioactive waste must therefore be subject to standards of safety.

Regulating safety is a national responsibility. However, radiation risks may transcend national borders, and international cooperation serves to promote and enhance safety globally by exchanging experience and by improving capabilities to control hazards, to prevent accidents, to respond to emergencies and to mitigate any harmful consequences.

States have an obligation of diligence and duty of care, and are expected to fulfil their national and international undertakings and obligations.

International safety standards provide support for States in meeting their obligations under general principles of international law, such as those relating to environmental protection. International safety standards also promote and assure confidence in safety and facilitate international commerce and trade.

A global nuclear safety regime is in place and is being continuously improved. IAEA safety standards, which support the implementation of binding international instruments and national safety infrastructures, are a cornerstone of this global regime. The IAEA safety standards constitute a useful tool for contracting parties to assess their performance under these international conventions.

THE IAEA SAFETY STANDARDS

The status of the IAEA safety standards derives from the IAEA's Statute, which authorizes the IAEA to establish or adopt, in consultation and, where appropriate, in collaboration with the competent organs of the United Nations and with the specialized agencies concerned, standards of safety for protection of health and minimization of danger to life and property, and to provide for their application.

With a view to ensuring the protection of people and the environment from harmful effects of ionizing radiation, the IAEA safety standards establish fundamental safety principles, requirements and measures to control the radiation exposure of people and the release of radioactive material to the environment, to restrict the likelihood of events that might lead to a loss of control over a nuclear reactor core, nuclear chain reaction, radioactive source or any other source of radiation, and to mitigate the consequences of such events if they were to occur. The standards apply to facilities and activities that give rise to radiation risks, including nuclear installations, the use of radiation and radioactive sources, the transport of radioactive material and the management of radioactive waste.

Safety measures and security measures[1] have in common the aim of protecting human life and health and the environment. Safety measures and security measures must be designed and implemented in an integrated manner so that security measures do not compromise safety and safety measures do not compromise security.

The IAEA safety standards reflect an international consensus on what constitutes a high level of safety for protecting people and the environment from harmful effects of ionizing radiation. They are issued in the IAEA Safety Standards Series, which has three categories (see Fig. 1).

---

[1] See also publications issued in the IAEA Nuclear Security Series.

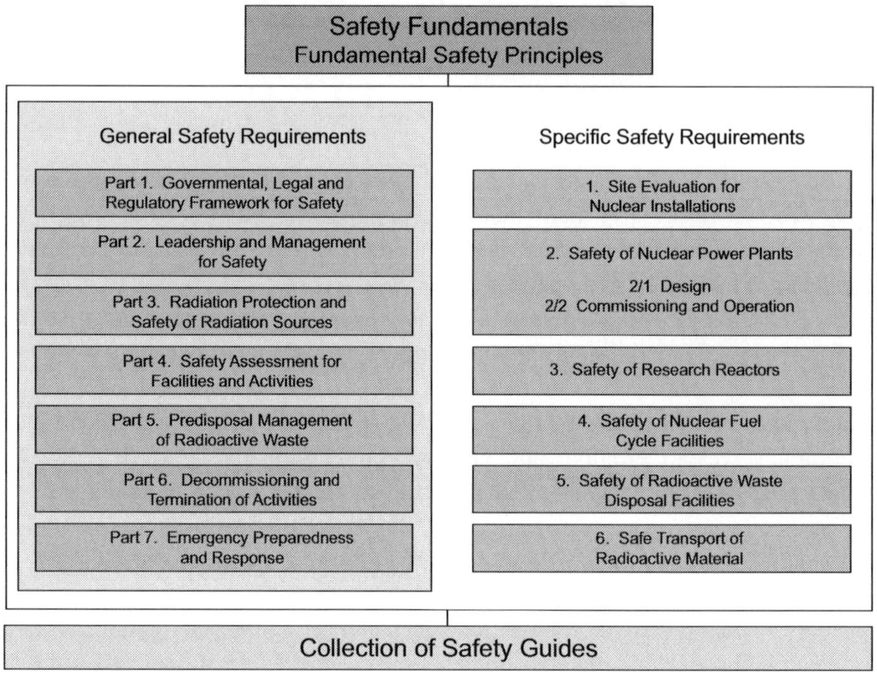

FIG. 1. The long term structure of the IAEA Safety Standards Series.

Safety Fundamentals

Safety Fundamentals present the fundamental safety objective and principles of protection and safety, and provide the basis for the safety requirements.

Safety Requirements

An integrated and consistent set of Safety Requirements establishes the requirements that must be met to ensure the protection of people and the environment, both now and in the future. The requirements are governed by the objective and principles of the Safety Fundamentals. If the requirements are not met, measures must be taken to reach or restore the required level of safety. The format and style of the requirements facilitate their use for the establishment, in a harmonized manner, of a national regulatory framework. Requirements, including numbered 'overarching' requirements, are expressed as 'shall' statements. Many requirements are not addressed to a specific party, the implication being that the appropriate parties are responsible for fulfilling them.

Safety Guides

Safety Guides provide recommendations and guidance on how to comply with the safety requirements, indicating an international consensus that it is necessary to take the measures recommended (or equivalent alternative measures). The Safety Guides present international good practices, and increasingly they reflect best practices, to help users striving to achieve high levels of safety. The recommendations provided in Safety Guides are expressed as 'should' statements.

APPLICATION OF THE IAEA SAFETY STANDARDS

The principal users of safety standards in IAEA Member States are regulatory bodies and other relevant national authorities. The IAEA safety standards are also used by co-sponsoring organizations and by many organizations that design, construct and operate nuclear facilities, as well as organizations involved in the use of radiation and radioactive sources.

The IAEA safety standards are applicable, as relevant, throughout the entire lifetime of all facilities and activities — existing and new — utilized for peaceful purposes and to protective actions to reduce existing radiation risks. They can be used by States as a reference for their national regulations in respect of facilities and activities.

The IAEA's Statute makes the safety standards binding on the IAEA in relation to its own operations and also on States in relation to IAEA assisted operations.

The IAEA safety standards also form the basis for the IAEA's safety review services, and they are used by the IAEA in support of competence building, including the development of educational curricula and training courses.

International conventions contain requirements similar to those in the IAEA safety standards and make them binding on contracting parties. The IAEA safety standards, supplemented by international conventions, industry standards and detailed national requirements, establish a consistent basis for protecting people and the environment. There will also be some special aspects of safety that need to be assessed at the national level. For example, many of the IAEA safety standards, in particular those addressing aspects of safety in planning or design, are intended to apply primarily to new facilities and activities. The requirements established in the IAEA safety standards might not be fully met at some existing facilities that were built to earlier standards. The way in which IAEA safety standards are to be applied to such facilities is a decision for individual States.

The scientific considerations underlying the IAEA safety standards provide an objective basis for decisions concerning safety; however, decision makers must also make informed judgements and must determine how best to balance the benefits of an action or an activity against the associated radiation risks and any other detrimental impacts to which it gives rise.

DEVELOPMENT PROCESS FOR THE IAEA SAFETY STANDARDS

The preparation and review of the safety standards involves the IAEA Secretariat and four safety standards committees, for nuclear safety (NUSSC), radiation safety (RASSC), the safety of radioactive waste (WASSC) and the safe transport of radioactive material (TRANSSC), and a Commission on Safety Standards (CSS) which oversees the IAEA safety standards programme (see Fig. 2).

All IAEA Member States may nominate experts for the safety standards committees and may provide comments on draft standards. The membership of the Commission on Safety Standards is appointed by the Director General and includes senior governmental officials having responsibility for establishing national standards.

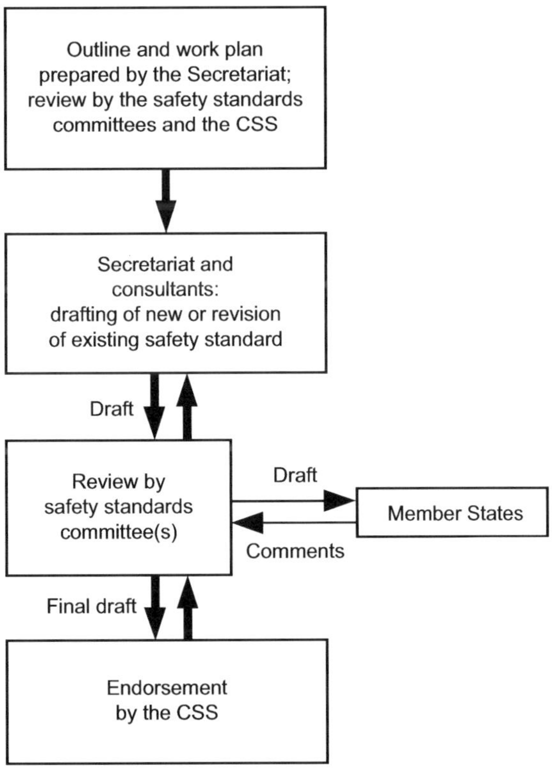

FIG. 2. The process for developing a new safety standard or revising an existing standard.

A management system has been established for the processes of planning, developing, reviewing, revising and establishing the IAEA safety standards. It articulates the mandate of the IAEA, the vision for the future application of the safety standards, policies and strategies, and corresponding functions and responsibilities.

INTERACTION WITH OTHER INTERNATIONAL ORGANIZATIONS

The findings of the United Nations Scientific Committee on the Effects of Atomic Radiation (UNSCEAR) and the recommendations of international expert bodies, notably the International Commission on Radiological Protection (ICRP), are taken into account in developing the IAEA safety standards. Some safety standards are developed in cooperation with other bodies in the United Nations system or other specialized agencies, including the Food and Agriculture Organization of the United Nations, the United Nations Environment Programme, the International Labour Organization, the OECD Nuclear Energy Agency, the Pan American Health Organization and the World Health Organization.

INTERPRETATION OF THE TEXT

Safety related terms are to be understood as defined in the IAEA Safety Glossary (see http://www-ns.iaea.org/standards/safety-glossary.htm). Otherwise, words are used with the spellings and meanings assigned to them in the latest edition of The Concise Oxford Dictionary. For Safety Guides, the English version of the text is the authoritative version.

The background and context of each standard in the IAEA Safety Standards Series and its objective, scope and structure are explained in Section 1, Introduction, of each publication.

Material for which there is no appropriate place in the body text (e.g. material that is subsidiary to or separate from the body text, is included in support of statements in the body text, or describes methods of calculation, procedures or limits and conditions) may be presented in appendices or annexes.

An appendix, if included, is considered to form an integral part of the safety standard. Material in an appendix has the same status as the body text, and the IAEA assumes authorship of it. Annexes and footnotes to the main text, if included, are used to provide practical examples or additional information or explanation. Annexes and footnotes are not integral parts of the main text. Annex material published by the IAEA is not necessarily issued under its authorship; material under other authorship may be presented in annexes to the safety standards. Extraneous material presented in annexes is excerpted and adapted as necessary to be generally useful.

CONTENTS

1. INTRODUCTION

BACKGROUND

1.1. This General Safety Requirements publication, IAEA Safety Standards Series No. GSR Part 3, Radiation Protection and Safety of Radiation Sources: International Basic Safety Standards (hereinafter referred to as 'these Standards'), is part of the IAEA Safety Standard Series, supersedes the International Basic Safety Standards for Protection against Ionizing Radiation and for the Safety of Radiation Sources (the BSS) issued in 1996[1]. Section 1 does not constitute a part of the requirements, but explains the context, concepts and principles for the requirements, which are established from Section 2 to Section 5, and in the Schedules.

1.2. Radioactivity is a natural phenomenon and natural sources of radiation are features of the environment. Radiation[2] and radioactive material may also be of artificial origin and have many beneficial applications, including uses in medicine, industry, agriculture and research as well as for nuclear power generation. The radiation risks to people and the environment that may arise from the use of radiation and radioactive material must be assessed and controlled through the application of standards of safety[3].

1.3. Exposure of tissues or organs to ionizing radiation can induce the death of cells on a scale that can be extensive enough to impair the function of the exposed tissue or organ. Effects of this type, which are called 'deterministic effects', are clinically observable in an individual only if the radiation dose exceeds a certain threshold. Above this threshold dose, a deterministic effect is more severe for a higher dose.

1.4. Exposure to radiation can also induce the non-lethal transformation of cells, which may still retain their capacity for cell division. The human body's immune system is very effective in detecting and destroying abnormal cells. However, there is a possibility that the non-lethal transformation of a cell could lead, after a latency period, to cancer in the individual exposed, if it is a somatic cell; or may lead to hereditary effects, if it is a germ cell. Such effects are called 'stochastic' effects. For the purposes of these Standards, it is assumed that the probability of the eventual occurrence of a stochastic effect is proportional to the dose received, with no threshold. The 'detriment-adjusted nominal risk coefficient of dose', which includes the risks of all cancers and hereditary effects, is 5% per sievert (Sv) [1]. This risk coefficient may need to be adjusted as new scientific knowledge becomes available.

1.5. The requirements established in these Standards are governed by the objectives, concepts and principles of the Fundamental Safety Principles [2]. These Standards draw upon information derived from the experience of States in applying the requirements of the previous International Basic Safety Standards[1], and from experience in many States in the use of radiation and nuclear techniques. These Standards draw upon extensive research and development work by national and international scientific and engineering organizations on the health effects of radiation exposure and on measures and techniques for the safe design and use of radiation sources. These Standards also take account of

---

[1] FOOD AND AGRICULTURE ORGANIZATION OF THE UNITED NATIONS, INTERNATIONAL ATOMIC ENERGY AGENCY, INTERNATIONAL LABOUR ORGANISATION, OECD NUCLEAR ENERGY AGENCY, PAN AMERICAN HEALTH ORGANIZATION, WORLD HEALTH ORGANIZATION, International Basic Safety Standards for Protection against Ionizing Radiation and for the Safety of Radiation Sources, Safety Series No. 115, IAEA, Vienna (1996).

[2] The term 'radiation' in this context means ionizing radiation.

[3] Obligations expressed as 'must' statements in Section 1 are quoted from the Fundamental Safety Principles [2].

the findings of the United Nations Committee on the Effects of Atomic Radiation (UNSCEAR) [4] and the Recommendations of the International Commission on Radiological Protection (ICRP) [1]. As scientific considerations are only part of the basis for making decisions on protection and safety, these Standards also address the use of value judgements relating to the management of risks.

The system of protection and safety

1.6. As stated in the Fundamental Safety Principles [2], "The fundamental safety objective is to protect people and the environment from harmful effects of ionizing radiation". This objective must be achieved without unduly limiting the operation of facilities or the conduct of activities that give rise to radiation risks[4]. Therefore, the system of protection and safety aims to assess, manage and control exposure to radiation so that radiation risks, including risks of health effects and risks to the environment, are reduced to the extent reasonably achievable.

1.7. These Standards are based on the following safety principles stated in the Fundamental Safety Principles [2]:

Principle 1: Responsibility for safety

> The prime responsibility for safety must rest with the person or organization responsible for facilities and activities[5] that give rise to radiation risks.

Principle 2: Role of government

> An effective legal and governmental framework for safety, including an independent regulatory body, must be established and sustained.

Principle 3: Leadership and management for safety

> Effective leadership and management for safety must be established and sustained in organizations concerned with, and facilities and activities that give rise to, radiation risks.

Principle 4: Justification of facilities and activities

> Facilities and activities that give rise to radiation risks must yield an overall benefit.

Principle 5: Optimization of protection

> Protection must be optimized to provide the highest level of safety that can reasonably be achieved.

---

[4] The term 'radiation risks' is used in a general sense to refer to:

- Detrimental health effects of radiation exposure (including the likelihood of such effects occurring).
- Any other safety related risks (including those to the environment) that might arise as a direct consequence of:
  - Exposure to radiation;
  - The presence of radioactive material (including radioactive waste) or its release to the environment;
  - A loss of control over a nuclear reactor core, nuclear chain reaction, radioactive source or any other source of radiation.

[5] The term 'facilities and activities' is a general term encompassing any human activity that may cause people to be exposed to radiation risks arising from naturally occurring or artificial sources. The term 'facilities' includes: nuclear facilities; irradiation installations; some mining and raw material processing facilities such as uranium mines; radioactive waste management facilities; and any other places where radioactive material is produced, processed, used, handled, stored or disposed of — or where radiation generators are installed — on such a scale that consideration of protection and safety is required. The term 'activities' includes: the production, use, import and export of radiation sources for industrial, research and medical purposes; the transport of radioactive material; the decommissioning of facilities; radioactive waste management activities such as the discharge of effluents; and some aspects of the remediation of sites affected by residues from past activities.

Principle 6: Limitation of risks to individuals

Measures for controlling radiation risks must ensure that no individual bears an unacceptable risk of harm.

Principle 7: Protection of present and future generations

People and the environment, present and future, must be protected against radiation risks.

Principle 8: Prevention of accidents

All practical efforts must be made to prevent and mitigate nuclear or radiation accidents.

Principle 9: Emergency preparedness and response

Arrangements must be made for emergency preparedness and response for nuclear or radiation incidents.

Principle 10: Protective actions to reduce existing or unregulated radiation risks

Protective actions to reduce existing or unregulated radiation risks must be justified and optimized.

The three general principles of radiation protection, which concern justification, optimization of protection and application of dose limits, are expressed in Safety Principles 4, 5, 6 and 10.

1.8. The prime responsibility for safety must rest with the person or organization responsible for facilities and activities that give rise to radiation risks [2]. Other parties also bear certain responsibilities. For instance, suppliers of radiation generators and radioactive sources have responsibilities in relation to the design and manufacture and operating instructions for their safe use. In the case of medical exposures, because of the medical setting in which such exposures occur, primary responsibility for protection and safety for patients lies with the health professional responsible for administration of the radiation dose, who is referred to in these Standards as the 'radiological medical practitioner'. Other types of health professionals may be involved in the preparation for, and the conduct of, radiological procedures, and each type has specific responsibilities, as established in these Standards.

1.9. A properly established governmental, legal and regulatory framework for safety provides for the regulation of facilities and activities that give rise to radiation risks. There is a hierarchy of responsibilities within this framework, from governments to regulatory bodies to the organizations responsible for and the persons engaged in activities involving radiation exposure. The government is responsible for the adoption within its national legal system of such legislation, regulations, and standards and measures as may be necessary to fulfil all its national and international obligations effectively, and for the establishment of an independent regulatory body. In some cases, more than one governmental organization may have the functions of a regulatory body for activities within their jurisdictions relating to the control of radiation and radioactive material.

1.10. Both the government and the regulatory body have important responsibilities in establishing the regulatory framework for protecting people and the environment from harmful effects of radiation, including establishing standards. These Standards require the government to ensure that there is coordination of government departments and agencies that have responsibilities for protection and safety, including the regulatory body, and departments and agencies concerned with public health, the environment, labour, mining, science and technology, agriculture and education. Standards have to be developed through consultation with those who are or could be required to apply them.

1.11. The government is also responsible for ensuring, as necessary, that provision is made for support services such as education and training, and technical services. If these services are not available within the State, other mechanisms to provide them may have to be considered. The regulatory body is responsible for carrying out its required regulatory functions, such as the establishment of requirements and guidelines, the authorization and inspection of facilities and activities, and the enforcement of legislative and regulatory provisions.

1.12. Leadership in safety matters has to be demonstrated at the highest levels in an organization, and safety has to be achieved and maintained by means of an effective management system. This system has to integrate all elements of management so that requirements for protection and safety are established and applied coherently with other requirements, including those for health, human performance, quality, protection of the environment and security, together with economic considerations. The application of the management system also has to ensure the promotion of a safety culture, the regular assessment of safety performance and the application of lessons learned from experience. Safety culture includes individual and collective commitment to safety on the part of the leadership, the management and personnel at all levels. The term 'management system' reflects and includes the concept of 'quality control' (controlling the quality of products) and its evolution through 'quality assurance' (the system for ensuring the quality of products) and 'quality management system' (the system for managing quality).

1.13. The operation of facilities or the conduct of activities that introduce a new source of radiation, that change exposures or that change the likelihood of exposures has to be justified in the sense that the detriments that may be caused are outweighed by the individual and societal benefits that are expected. The comparison of detriments and benefits often goes beyond the consideration of protection and safety and also involves the consideration of economic, societal and environmental factors.

1.14. The application of the justification principle to medical exposures requires a special approach. As an overarching justification of medical exposures, it is accepted that the use of radiation in medicine does more good than harm. However, at the next level, there is a need for generic justification, to be carried out by the health authority in conjunction with appropriate professional bodies, of a given radiological procedure. This applies to the justification of new technologies and techniques as they evolve. For the final level of justification, the application of the radiological procedure to a given individual has to be considered. The specific objectives of the exposure, the clinical circumstances and the characteristics of the individual involved have to be taken into account through referral criteria developed by professional bodies and the health authority.

1.15. The optimization of protection and safety, when applied to the exposure of workers and of members of the public, and of 'carers and comforters' of patients undergoing radiological procedures, is a process for ensuring that the magnitude and likelihood of exposures and the number of individuals exposed are as low as reasonably achievable, with economic, societal and environmental factors taken into account. This means that the level of protection would be the best possible under the prevailing circumstances. Optimization is a prospective and iterative process that requires both qualitative and quantitative judgements to be made.

1.16. As is the case with justification, the application of the optimization principle to the medical exposure of patients and to that of volunteers as part of a programme of biomedical research requires a special approach. Too low a radiation dose could be as bad as too high a radiation dose, in that the consequence could be that a cancer is not cured or the images taken are not of suitable diagnostic quality. It is of paramount importance that the medical exposure leads to the required outcome.

1.17. For planned exposure situations, exposures and risks are subject to control to ensure that the specified dose limits for occupational exposure and those for public exposure are not exceeded, and optimization is applied to attain the desired level of protection and safety.

1.18. All practical efforts must be made to prevent and mitigate nuclear or radiation accidents. The most harmful consequences arising from facilities and activities have come from the loss of control over a nuclear reactor core, nuclear chain reaction, radioactive source or other source of radiation. Consequently, to ensure that the likelihood of an accident having harmful consequences is extremely low, measures have to be taken:

- To prevent the occurrence of failures or abnormal conditions (including breaches of security) that could lead to such a loss of control;
- To prevent the escalation of any such failures or abnormal conditions that do occur;
- To prevent the loss of, or the loss of control over, a radioactive source or other source of radiation.

1.19. Arrangements must be made for emergency preparedness and response for nuclear or radiation incidents. The primary goals of preparedness and response for a nuclear or radiation emergency are:

- To ensure that arrangements are in place for an effective response at the scene and, as appropriate, at the local, regional, national and international level;
- To ensure that, for reasonably foreseeable incidents, radiation risks would be minor;
- To take practical measures to mitigate any consequences for human life and health and the environment, for any incidents that do occur.

Types of exposure situation

1.20. For the purpose of establishing practical requirements for protection and safety, these Standards distinguish between three different types of exposure situation: planned exposure situations, emergency exposure situations and existing exposure situations [1]. Together, these three types of exposure situation cover all situations of exposure to which these Standards apply.

(i) A *planned exposure situation* is a situation of exposure that arises from the planned operation of a source or from a planned activity that results in an exposure from a source. Since provision for protection and safety can be made before embarking on the activity concerned, the associated exposures and their likelihood of occurrence can be restricted from the outset. The primary means of controlling exposure in planned exposure situations are by good design of facilities, equipment and operating procedures and by training. In planned exposure situations, exposure at some level can be expected to occur. If exposure is not expected to occur with certainty, but could result from an accident or from an event or a sequence of events that may occur but is not certain to occur, this is referred to as 'potential exposure'.

(ii) An *emergency exposure situation* is a situation of exposure that arises as a result of an accident, a malicious act, or any other unexpected event, and requires prompt action in order to avoid or to reduce adverse consequences. Preventive actions and mitigatory actions have to be considered before an emergency exposure situation arises. However, once an emergency exposure situation actually occurs, exposures can be reduced only by implementing protective actions.

(iii) An *existing exposure situation* is a situation of exposure which already exists when a decision on the need for control needs to be taken. Existing exposure situations include situations of exposure to natural background radiation. They also include situations of exposure due to residual radioactive material that derives from past practices that were not subject to regulatory control or that remains after an emergency exposure situation.

If an event or a sequence of events that has been considered in the assessment of potential exposure does actually occur, it may be treated either as a planned exposure situation or, if an emergency is declared, as an emergency exposure situation.

1.21. The descriptions that are given in para. 1.20 of the three types of exposure situation are not always sufficient to determine unequivocally which type of exposure situation applies for particular circumstances. For instance, the transition from an emergency exposure situation to an existing exposure situation may occur progressively over time; and some exposures due to natural sources may have some characteristics of both planned exposure situations and existing exposure situations. In these Standards, the most appropriate type of exposure situation for particular circumstances has been determined by taking practical considerations into account. For the purposes of these Standards, the exposure of aircrew to cosmic radiation is considered under existing exposure situations in Section 5. The exposure of space crew to cosmic radiation presents exceptional circumstances and these are addressed separately in Section 5.

Dose constraints and reference levels

1.22. Dose constraints and reference levels are used for optimization of protection and safety, the intended outcome of which is that all exposures are controlled to levels that are as low as reasonably achievable, economic, societal and environmental factors being taken into account. Dose constraints are applied to occupational exposure and to public exposure in planned exposure situations. Dose constraints are set separately for each source under control and they serve as boundary conditions in defining the range of options for the purposes of optimization. Dose constraints are not dose limits; exceeding a dose constraint does not represent non-compliance with regulatory requirements, but it could result in follow-up actions.

1.23. While the objectives of the use of dose constraints for controlling occupational exposure and public exposure are similar, the dose constraints are applied in different ways. For occupational exposure, the dose constraint is a tool to be established and used in the optimization of protection and safety by the person or organization responsible for a facility or activity. For public exposure in planned exposure situations, the government or the regulatory body ensures the establishment or approval of dose constraints, taking into account the characteristics of the site and of the facility or activity, the scenarios for exposure and the views of interested parties. After exposures have occurred, the dose constraint may be used as a benchmark for assessing the suitability of the optimized strategy for protection and safety (referred to as the protection strategy) that has been implemented and for making adjustments as necessary. The setting of the dose constraint needs to be considered in conjunction with other health and safety provisions and the technology available.

1.24. Reference levels are used for optimization of protection and safety in emergency exposure situations and in existing exposure situations. They are established or approved by the government, the regulatory body or another relevant authority. For occupational exposure and public exposure in emergency exposure situations and in existing exposure situations, a reference level serves as a boundary condition in defining the range of options for the purposes of optimization in implementing protective actions. The reference level represents the level of dose or the level of risk above which it is judged to be inappropriate to plan to allow exposures to occur, and below which the optimization of protection and safety is implemented. The value chosen for the reference level will depend upon the prevailing circumstances for the exposures under consideration. The optimized protection strategies are intended to keep doses below the reference level. When an emergency exposure situation has arisen or an existing exposure situation has been identified, actual exposures could be above or below the reference level. The reference level would be used as a benchmark for judging whether further protective measures are necessary and, if so, in prioritizing their application. Optimization is to be applied in emergency exposure situations and in existing exposure situations, even if the doses initially received are below the reference level.

1.25. The ICRP recommends a range of dose spanning two orders of magnitude within which the value of a dose constraint or reference level would usually be chosen [1]. At the lower end of this range, the dose constraint or reference level represents an increase, of up to about 1 mSv, over the dose received in a year from exposure due to naturally occurring radiation sources[6]. It would be used when persons are exposed to radiation from a source that yields little or no benefit for them, but which may benefit society in general. This would be the case, for instance, in establishing dose constraints for public exposure in planned exposure situations.

1.26. Dose constraints or reference levels of 1–20 mSv would be used when the exposure situation, but not necessarily the exposure itself, usually benefits individuals. This would be the case, for instance, when establishing dose constraints for occupational exposure in planned exposure situations or reference levels for exposure of a member of the public in existing exposure situations.

1.27. Reference levels of 20–100 mSv would be used where individuals are exposed to radiation from sources that are not under control or where actions to reduce doses would be disproportionately disruptive. This would be the case, for instance, in establishing reference levels for the residual dose after a nuclear or radiation emergency. Any situation that resulted in a dose of greater than 100 mSv being incurred acutely or in one year would be considered unacceptable, except under the circumstances relating to exposure of emergency workers that are addressed specifically in these Standards.

1.28. The selection of the value for the dose constraint or the reference level would be based on the characteristics of the exposure situation, including:

(i) The nature of the exposure and the practicability of reducing or preventing the exposure;

(ii) The expected benefits of the exposure for individuals and society, or the benefits of avoiding preventive actions or protective actions that would be detrimental to living conditions, as well as other societal criteria relating to the management of the exposure situation;

(iii) National or regional factors, together with a consideration of international guidance and good practice elsewhere.

1.29. The system of protection and safety required by these Standards includes criteria for protection against exposure due to radon which are based on the average level of risk to a population with typical but various smoking habits. Because of the synergistic effects of smoking and exposure due to radon, the absolute risk of lung cancer resulting from unit dose due to radon for people who are smokers is substantially greater than for those who have never smoked [4, 5, 6]. Information provided to people on the risks associated with exposure due to radon needs to highlight this increased risk for smokers.

1.30. Dose constraints are used in optimization of protection and safety for ‘carers and comforters’ and for volunteers subject to exposure as part of a programme of biomedical research. Dose constraints are not applicable to the exposure of patients in radiological procedures for the purposes of medical diagnosis or treatment.

1.31. In X ray medical imaging, image guided interventional procedures and diagnostic nuclear medicine, a diagnostic reference level is used to indicate the need for an investigation. Periodic

---

[6] According to the United Nations Scientific Committee on the Effects of Atomic Radiation (UNSCEAR) [3], the worldwide average annual radiation dose from exposure due to naturally occurring radiation sources, including radon, is 2.4 mSv. In any large population, about 65% would be expected to have annual doses of between 1 and 3 mSv. About 25% of the population would be expected to have annual doses of less than 1 mSv, and about 10% would be expected to have annual doses greater than 3 mSv.

assessments are performed of typical doses or activity of the radiopharmaceuticals administered in a medical facility. If comparison with established diagnostic reference levels shows that the typical doses or activity of the radiopharmaceuticals administered are either too high or unusually low, a local review is to be initiated to ascertain whether protection and safety has been optimized and whether any corrective action is required.

Protection of the environment

1.32. In a global and long term perspective, protection of people and the environment against radiation risks associated with the operation of facilities and the conduct of activities — and in particular, protection against such risks that may transcend national borders and may persist for long periods of time — is important to achieving equitable and sustainable development.

1.33. The system of protection and safety required by these Standards generally provides for appropriate protection of the environment from harmful effects of radiation. Nevertheless, international trends in this field show an increasing awareness of the vulnerability of the environment. Trends also indicate the need to be able to demonstrate (rather than to assume) that the environment is protected against effects of industrial pollutants, including radionuclides, in a wider range of environmental situations, irrespective of any human connection. This is usually accomplished by means of an environmental assessment that identifies the target(s), defines the appropriate criteria for protection, assesses the impacts and compares the expected results of the available protection options. Methods and criteria for such assessments are being developed and will continue to evolve.

1.34. Radiological impacts in a particular environment constitute only one type of impact and, in most cases, may not be the dominant impact of a particular facility or activity. Furthermore, the assessment of impacts on the environment needs to be viewed in an integrated manner with other features of the system of protection and safety to establish the requirements applicable to a particular source. Since there are complex interrelations, the approach to the protection of people and the environment is not limited to the prevention of radiological effects on humans and on other species. When establishing regulations, an integrated perspective has to be adopted to ensure the sustainability, now and in the future, of agriculture, forestry, fisheries and tourism, and of the use of natural resources. Such an integrated perspective also has to take into account the need to prevent unauthorized acts with potential consequences for and via the environment, including, for example, illicit dumping of radioactive material and the abandonment of radiation sources. Consideration also needs to be given to the potential for buildup and accumulation of long lived radionuclides released to the environment.

1.35. These Standards are designed to identify the protection of the environment as an issue necessitating assessment, while allowing for flexibility in incorporating into decision making processes the results of environmental assessments that are commensurate with the radiation risks.

Interfaces between safety and security

1.36. Safety measures and security measures have in common the aim of protecting human life and health and the environment. In addition, safety measures and security measures must be designed and implemented in an integrated manner so that security measures do not compromise safety and safety measures do not compromise security.

1.37. Security infrastructure and safety infrastructure need to be developed, as far as possible, in a well coordinated manner. All organizations involved need to be made aware of the commonalities and the differences between safety and security so as to be able to factor both into development plans. The synergies between safety and security have to be developed so that safety and security complement and enhance one another.

OBJECTIVE

1.38. These Standards establish requirements for the protection of people and the environment from harmful effects of ionizing radiation and for the safety of radiation sources.

SCOPE

1.39. These Standards apply for protection against ionizing radiation only, which includes gamma rays, X rays and particles such as beta particles, neutrons, protons, alpha particles and heavier ions. While these Standards do not specifically address the control of non-radiological aspects of health, safety and the environment, these aspects also need to be considered. Protection from harmful effects of non-ionizing radiation is outside the scope of these Standards.

1.40. These Standards are intended primarily for use by governments and regulatory bodies. Requirements also apply to principal parties and other parties as specified in Section 2, health authorities, professional bodies and service providers such as technical support organizations.

1.41. These Standards do not deal with security measures. The IAEA issues recommendations on nuclear security in the IAEA Nuclear Security Series.

1.42. These Standards apply to all situations involving radiation exposure that is amenable to control. Exposures deemed to be unamenable to control are excluded from the scope of these Standards[7].

1.43. These Standards establish requirements to be fulfilled in all facilities and activities giving rise to radiation risks. For certain facilities and activities, such as nuclear installations, radioactive waste management facilities and the transport of radioactive material, other safety requirements, complementary to these Standards, also apply. The IAEA issues Safety Guides to assist in the application of these Standards.

1.44. These Standards apply to the three categories of exposure: occupational exposure, public exposure and medical exposure.

1.45. These Standards apply to human activities involving radiation exposure that are:

(i) Carried out in a State which chooses to adopt these Standards or which requests any of the Sponsoring Organizations to provide for the application of these Standards;
(ii) Undertaken by States with the assistance of the Food and Agriculture Organization of the United Nations, the International Atomic Energy Agency, the International Labour Organization, the Pan American Health Organization, the United Nations Environment Programme or the World Health Organization, in the light of relevant national rules and regulations;
(iii) Carried out by the IAEA or involving the use of materials, services, equipment, facilities and non-published information made available by the IAEA or at its request or under its control or supervision; or
(iv) Carried out under any bilateral or multilateral arrangement whereby the parties request the IAEA to provide for the application of these Standards.

1.46. Quantities and units used in these Standards are in accordance with the recommendations of the International Commission on Radiation Units and Measurements (ICRU) [7].

---

[7] It is generally accepted, for example, that it is not feasible to control ^{40}K in the body or cosmic radiation at the surface of the Earth.

STRUCTURE

1.47. The requirements of these Standards are grouped into requirements applicable for all exposure situations and separate requirements for planned exposure situations, emergency exposure situations and existing exposure situations. For each of the three types of exposure situation, the requirements are further grouped into requirements for occupational exposure, public exposure and (for planned exposure situations) medical exposure.

1.48. The requirements established by these Standards, both numbered 'overarching' requirements in bold with titles and other requirements, are expressed as 'shall' statements. Each individual overarching requirement is followed by associated requirements.

1.49. Section 2 sets out the requirements that apply generally for all exposure situations and for all three categories of exposure (occupational exposure, public exposure and medical exposure). These requirements include the assignment of responsibilities to the government, the regulatory body, and principal parties and other parties with respect to the implementation of a protection and safety programme and a management system, the promotion of a safety culture and the consideration of human factors.

1.50. Section 3 sets out the requirements — in addition to those of Section 2 — for planned exposure situations. Section 3 includes requirements applicable to all three categories of exposure, requirements for the safety of sources, and separate requirements in respect of occupational exposure, public exposure and medical exposure.

1.51. Section 4 sets out the requirements — in addition to those of Section 2 — for emergency exposure situations. Section 4 includes requirements in respect of public exposure and occupational exposure (exposure of emergency workers) in emergency exposure situations. It also includes requirements on the transition from an emergency exposure situation to an existing exposure situation.

1.52. Section 5 sets out the requirements — in addition to those of Section 2 — for existing exposure situations. Section 5 includes requirements in respect of public exposure and occupational exposure in existing exposure situations. It includes requirements in respect of remediation of sites and habitation in areas with residual radioactive material, radon in homes and in workplaces, radionuclides in commodities, and exposure of aircrew and space crew.

1.53. The organization of the requirements in these Standards for the relevant categories of exposure in each type of exposure situation is as shown in Table 1. General requirements for all exposure situations are given in Section 2, and requirements for different exposure situations are given in Sections 3, 4 and 5. Thus, for any particular facility or activity, more than one section of these Standards will be relevant, as illustrated by the following examples:

(i) The requirements for the regulatory body given in Section 2 are applicable for all exposure situations and all categories of exposure. They provide the regulatory framework within which persons or organizations responsible for facilities and activities have to comply with the requirements placed on them. These requirements thus establish the general regulatory responsibilities of the regulatory body. Any further requirements on the regulatory body that apply for one type of exposure situation are given in Sections 3, 4 and 5. These requirements are in addition to the requirements given in Section 2.

(ii) Persons or organizations responsible for a medical facility in which radiation generators or radioactive sources are used are subject to the requirements given in Section 2 for all exposure situations and all categories of exposure, and also to those requirements given in Section 3 that are common to all planned exposure situations (paras 3.5–3.67). In addition, they are subject to the separate requirements given in Section 3 for occupational exposure (such as exposure of medical staff operating medical devices that emit radiation) (paras 3.68–3.116), public exposure (such as exposure in rooms adjacent to rooms containing equipment that generates radiation) (paras 3.117–3.143) and medical exposure (such as exposure of patients) (paras 3.144–3.184).

TABLE 1. ORGANIZATION OF THE REQUIREMENTS OF THESE STANDARDS

	Occupational exposure	Public exposure	Medical exposure
Planned exposure situations	Section 2; Section 3: paras 3.5–3.67 and paras 3.68–3.116	Section 2; Section 3: paras 3.5–3.67 and paras 3.117–3.143	Section 2; Section 3: paras 3.5–3.67 and paras 3.144–3.184
Emergency exposure situations	Section 2; Section 4	Section 2; Section 4	*Not applicable*
Existing exposure situations	Section 2; Section 5	Section 2; Section 5	*Not applicable*

1.54. Four schedules provide numerical values in support of the requirements, covering exemption and clearance, categorization of sealed sources, dose limits for planned exposure situations and criteria for use in emergency preparedness and response.

1.55. Definitions of terms used are included in these Standards.

2. GENERAL REQUIREMENTS FOR PROTECTION AND SAFETY

DEFINITIONS

2.1. Terms used have the meanings given under Definitions.

INTERPRETATION

2.2. Except as specifically authorized by the statutory governing body of a relevant sponsoring organization, no interpretation of these Standards by any officer or employee of the sponsoring organization other than a written interpretation by the Director General of the sponsoring organization will be binding on the sponsoring organization.

RESOLUTION OF CONFLICTS

2.3. The requirements of these Standards are in addition to and not in place of other applicable requirements, such as those of relevant binding conventions and national regulations.

2.4. In cases of conflict between the requirements of these Standards and other applicable requirements, the government or the regulatory body, as appropriate, shall determine which requirements are to be enforced.

2.5. Nothing in these Standards shall be construed as restricting any actions that may otherwise be necessary for protection and safety or as relieving the parties referred to in paras 2.40 and 2.41 from complying with applicable laws and regulations.

ENTRY INTO FORCE

2.6. These Standards shall enter into force one year after the date of their adoption or acknowledgement, as appropriate, by the relevant Sponsoring Organization.

2.7. If a State decides to adopt these Standards, these Standards shall come into force at the time indicated in the formal adoption by that State.

APPLICATION OF THE PRINCIPLES OF RADIATION PROTECTION

Requirement 1: Application of the principles of radiation protection

Parties with responsibilities for protection and safety shall ensure that the principles of radiation protection are applied for all exposure situations.

2.8. For planned exposure situations, each party with responsibilities for protection and safety shall ensure, when relevant requirements apply to that party, that no practice is undertaken unless it is justified.

2.9. For emergency exposure situations and existing exposure situations, each party with responsibilities for protection and safety shall ensure, when relevant requirements apply to that party, that protective actions or remedial actions are justified and are undertaken in such a way as to achieve the objectives set out in a protection strategy.

2.10. For all exposure situations, each party with responsibilities for protection and safety shall ensure, when relevant requirements apply to that party, that protection and safety is optimized.[8]

2.11. For planned exposure situations other than for medical exposure, each party with responsibilities for protection and safety shall ensure that, when relevant requirements apply to that party, specified dose limits are not exceeded.

2.12. The application of the requirements for the system of protection and safety shall be commensurate with the radiation risks associated with the exposure situation.

RESPONSIBILITIES OF THE GOVERNMENT[9]

Requirement 2: Establishment of a legal and regulatory framework

The government shall establish and maintain a legal and regulatory framework for protection and safety and shall establish an effectively independent regulatory body with specified responsibilities and functions.

2.13. The government shall establish and maintain an appropriate and effective legal and regulatory framework for protection and safety in all exposure situations[10]. This framework shall encompass both the assignment and the discharge of governmental responsibilities, and the regulatory control of facilities and activities that give rise to radiation risks. The framework shall allow for the fulfilment of international obligations.

2.14. The government shall ensure that adequate arrangements are in place for the protection of people and the environment, both now and in the future, against harmful effects of ionizing radiation, without unduly limiting the operation of facilities or the conduct of activities that give rise to radiation risks. This shall include arrangements for the protection of people of present and future generations and populations remote from present facilities and activities.

2.15. The government shall establish legislation that, among other things:

(a) Provides the statutory basis for requirements for protection and safety for all exposure situations;
(b) Specifies that the prime responsibility for protection and safety rests with the person or organization responsible for facilities and activities that give rise to radiation risks;
(c) Specifies the scope of its applicability;
(d) Establishes and provides for maintaining an independent regulatory body with clearly specified functions and responsibilities for the regulation of protection and safety;
(e) Provides for coordination between authorities with responsibilities relevant to protection and safety for all exposure situations.

---

[8] 'Protection and safety is optimized' means that optimization of protection and safety has been applied and the result of that process has been implemented.

[9] States have different legal structures, and therefore the term 'government' as used in the IAEA safety standards is to be understood in a broad sense, and is accordingly interchangeable here with the term 'State'.

[10] Requirements on the governmental, legal and regulatory framework for safety of facilities and activities are established in Ref. [8].

2.16. The government shall ensure that the regulatory body is effectively independent, in making decisions relating to protection and safety, of persons and organizations using or promoting the use of radiation and radioactive material, so that it is free from any undue influence by interested parties and from any conflicts of interest, and that it has functional separation from entities having responsibilities or interests that could unduly influence its decision making.

2.17. The government shall ensure that the regulatory body has the legal authority, competence and resources necessary to fulfil its statutory functions and responsibilities.

2.18. The government shall ensure that a graded approach is taken to the regulatory control of radiation exposure, so that the application of regulatory requirements is commensurate with the radiation risks associated with the exposure situation.

2.19. The government shall establish mechanisms to ensure that:

(a) The activities of the regulatory body are coordinated with those of other governmental authorities, in accordance with para. 2.15(e), and with national and international organizations that have related responsibilities;
(b) Interested parties are involved as appropriate in regulatory decision making processes or regulatory decision aiding processes.

2.20. The government shall ensure that arrangements are in place at the national level for making decisions relating to protection and safety that fall outside the authority of the regulatory body.

2.21. The government shall ensure that requirements are established for:

(a) education, training, qualification and competence in protection and safety of all persons engaged in activities relevant to protection and safety;
(b) the formal recognition[11] of qualified experts;
(c) the competence of organizations that have responsibilities relating to protection and safety.

2.22. The government shall ensure that arrangements are in place for the provision of the education and training services required for building and maintaining the competence of persons and organizations that have responsibilities relating to protection and safety.

2.23. The government shall ensure that arrangements are in place for the provision of technical services relating to protection and safety, such as services for personal dosimetry, environmental monitoring and the calibration of monitoring and measuring equipment.

2.24. The government shall ensure that arrangements are in place for the safe decommissioning of facilities [9], the safe management of radioactive waste [10, 11] and the safe management of spent fuel.

2.25. The government shall ensure that the transport of radioactive material is in accordance with the IAEA Regulations for the Safe Transport of Radioactive Material [12] and with any applicable international conventions, taking into consideration other internationally endorsed standards and recommendations derived from these IAEA Regulations.[12]

---

[11] 'Formal recognition' means documented acknowledgment by the relevant authority that a person has the qualifications and expertise required for the responsibilities that he or she will bear in the conduct of the authorized activity.

[12] Additional measures are taken for security in the transport of radioactive material. The IAEA issues guidance on security in the transport of radioactive material in the IAEA Nuclear Security Series of publications.

2.26. The government shall ensure that arrangements are in place for regaining control over radioactive sources that have been abandoned, lost, misplaced, stolen or otherwise transferred without proper authorization.

2.27. The government shall ensure that infrastructural arrangements are in place for the interfaces between safety and the security of radioactive sources.

2.28. In establishing the legal and regulatory framework for protection and safety, the government:

(a) shall fulfil its respective international obligations;
(b) shall allow for participation in relevant international arrangements, including international peer reviews;
(c) shall promote international cooperation to enhance safety globally.

RESPONSIBILITIES OF THE REGULATORY BODY

Requirement 3: Responsibilities of the regulatory body

The regulatory body shall establish or adopt regulations and guides for protection and safety and shall establish a system to ensure their implementation.

2.29. The regulatory body shall establish requirements for the application of the principles of radiation protection specified in paras 2.8–2.12 for all exposure situations and shall establish or adopt regulations and guides for protection and safety.

2.30. The regulatory body shall establish a regulatory system for protection and safety that includes [8]:

(a) Notification and authorization;
(b) Review and assessment of facilities and activities;
(c) Inspection of facilities and activities;
(d) Enforcement of regulatory requirements;
(e) The regulatory functions relevant to emergency exposure situations and existing exposure situations;
(f) Provision of information to, and consultation with, parties affected by its decisions and, as appropriate, the public and other interested parties.

2.31. The regulatory body shall adopt a graded approach to the implementation of the system of protection and safety, such that the application of regulatory requirements is commensurate with the radiation risks associated with the exposure situation.

2.32. The regulatory body shall ensure the application of the requirements for education, training, qualification and competence in protection and safety of all persons engaged in activities relevant to protection and safety.

2.33. The regulatory body shall ensure that mechanisms are in place for the timely dissemination of information to relevant parties, such as suppliers and users of sources, on lessons learned for protection and safety from regulatory experience and operating experience, and from incidents and accidents and the related findings. The mechanisms established shall, as appropriate, be used to provide relevant information to other relevant organizations at the national and international level.

2.34. The regulatory body, in conjunction with other competent authorities, shall adopt specific acceptance criteria and performance criteria, through regulation or by the application of published standards, for any manufactured or constructed source, device, equipment or facility that, in use, has implications for protection and safety.

2.35. The regulatory body shall make provision for establishing, maintaining and retrieving adequate records relating to facilities and activities. These records shall include:

- Registers of sealed sources and radiation generators[13];
- Records of doses from occupational exposure;
- Records relating to the safety of facilities and activities;
- Records that might be necessary for the shutdown and decommissioning or closure of facilities;
- Records of events, including non-routine releases of radioactive material to the environment;
- Inventories of radioactive waste and of spent fuel.

2.36. The regulatory body shall establish mechanisms for communication and discussion that involve professional and constructive interactions with relevant parties for all protection and safety related issues.

2.37. The regulatory body, in consultation with the health authority, shall ensure that provisions are in place for ensuring protection and safety in the handling of deceased persons or human remains that are known to contain sealed or unsealed radioactive sources, either as a result of radiological procedures for medical treatment of patients or as a consequence of an emergency.

2.38. The regulatory body shall establish, implement, assess and strive to continually improve a management system that is aligned with the goals of the regulatory body and that contributes to the achievement of those goals.

RESPONSIBILITIES FOR PROTECTION AND SAFETY

Requirement 4: Responsibilities for protection and safety

The person or organization responsible for facilities and activities that give rise to radiation risks shall have the prime responsibility for protection and safety. Other parties shall have specified responsibilities for protection and safety.

2.39. The person or organization responsible for any facility or activity that gives rise to radiation risks shall have the prime responsibility for protection and safety, which cannot be delegated.

2.40. The principal parties responsible for protection and safety are:

(a) Registrants or licensees, or the person or organization responsible for facilities and activities for which notification only is required;
(b) Employers, in relation to occupational exposure;
(c) Radiological medical practitioners, in relation to medical exposure;

---

[13] The regulatory body specifies which sources are to be included in the registers and inventories, with due consideration given to the associated risks.

(d) Those persons or organizations designated to deal with emergency exposure situations or existing exposure situations.

2.41. Other parties shall have specified responsibilities in relation to protection and safety. These other parties include:

(a) Suppliers of sources, providers of equipment and software, and providers of consumer products;
(b) Radiation protection officers;
(c) Referring medical practitioners;
(d) Medical physicists;
(e) Medical radiation technologists;
(f) Qualified experts or any other party to whom a principal party has assigned specific responsibilities;
(g) Workers other than workers listed in (a)–(f);
(h) Ethics committees.

2.42. The relevant principal parties shall establish and implement a protection and safety programme that is appropriate for the exposure situation. The protection and safety programme:

(a) Shall adopt objectives for protection and safety in accordance with the requirements of these Standards;
(b) Shall apply measures for protection and safety that are commensurate with the radiation risks associated with the exposure situation and that are adequate to ensure compliance with the requirements of these Standards.

2.43. The relevant principal parties shall ensure that, in the implementation of the protection and safety programme:

(a) The measures and resources necessary for achieving the objectives for protection and safety have been determined and are duly provided;
(b) The programme is periodically reviewed to assess its effectiveness and its continued fitness for purpose;
(c) Any failures or shortcomings in protection and safety are identified and corrected, and steps are taken to prevent their recurrence;
(d) Arrangements are made to consult with relevant interested parties;
(e) Appropriate records are maintained.

2.44. The relevant principal parties and other parties having specified responsibilities in relation to protection and safety shall ensure that all personnel engaged in activities relevant to protection and safety have appropriate education, training and qualification so that they understand their responsibilities and can perform their duties competently, with appropriate judgement and in accordance with procedures.

2.45. The relevant principal parties shall permit access by authorized representatives of the regulatory body to carry out inspections of their facilities and activities and of their protection and safety records, and shall cooperate in the conduct of inspections.

2.46. The relevant principal parties shall ensure that qualified experts are identified and consulted as necessary on the proper observance of these Standards.

MANAGEMENT REQUIREMENTS

Requirement 5: Management for protection and safety

The principal parties shall ensure that protection and safety is effectively integrated into the overall management system of the organizations for which they are responsible.

Protection and safety elements of the management system

2.47. The principal parties shall demonstrate commitment to protection and safety at the highest levels within the organizations for which they are responsible.

2.48. The principal parties shall ensure that the management system[14] is designed and implemented to enhance protection and safety by:

(a) Applying the requirements for protection and safety coherently with other requirements, including requirements for operational performance, and coherently with guidelines for security;
(b) Describing the planned and systematic actions necessary to provide adequate confidence that the requirements for protection and safety are fulfilled;
(c) Ensuring that protection and safety is not compromised by other requirements;
(d) Providing for the regular assessment of performance for protection and safety and the application of lessons learned from experience;
(e) Promoting safety culture.

2.49. The principal parties shall ensure that protection and safety elements of the management system are commensurate with the complexity of and the radiation risks associated with the activity.

2.50. The principal parties shall be able to demonstrate the effective fulfilment of the requirements for the protection and safety in the management system.

Safety culture

2.51. The principal parties shall promote and maintain a safety culture by:

(a) Promoting individual and collective commitment to protection and safety at all levels of the organization;
(b) Ensuring a common understanding of the key aspects of safety culture within the organization;
(c) Providing the means by which the organization supports individuals and teams in carrying out their tasks safely and successfully, with account taken of the interactions between individuals, technology and the organization;
(d) Encouraging the participation of workers and their representatives and other relevant persons in the development and implementation of policies, rules and procedures dealing with protection and safety;
(e) Ensuring accountability of the organization and of individuals at all levels for protection and safety;
(f) Encouraging open communication with regard to protection and safety within the organization and with relevant parties, as appropriate;
(g) Encouraging a questioning and learning attitude and discouraging complacency with regard to protection and safety;
(h) Providing means by which the organization continually seeks to develop and strengthen its safety culture.

---

[14] Requirements on the management systems for facilities and activities are established in Ref. [13].

Human factors

2.52. The principal parties and other parties having specified responsibilities in relation to protection and safety, as appropriate, shall take into account human factors and shall support good performance and good practices to prevent human and organizational failures, by ensuring among other things that:

(a) Sound ergonomic principles are followed in the design of equipment and the development of operating procedures, so as to facilitate the safe operation and use of equipment, to minimize the possibility that operator errors will lead to accidents, and to reduce the possibility that indications of normal conditions and abnormal conditions will be misinterpreted;
(b) Appropriate equipment, safety systems and procedural requirements are provided and other necessary provisions are made:
   (i) To reduce, as far as practicable, the possibility that human error or inadvertent action could give rise to accidents or other incidents leading to the exposure of any person;
   (ii) To provide means for detecting human errors and for correcting them or compensating for them;
   (iii) To facilitate protective actions and corrective actions in the event of failures of safety systems or failures of protective measures.

3. PLANNED EXPOSURE SITUATIONS

SCOPE

3.1. The requirements for planned exposure situations apply to the following practices:

(a) The production, supply and transport of radioactive material and of devices that contain radioactive material, including sealed sources and unsealed sources, and of consumer products;
(b) The production and supply of devices that generate radiation, including linear accelerators, cyclotrons, and fixed and mobile radiography equipment;
(c) The generation of nuclear power, including any activities within the nuclear fuel cycle that involve or that could involve exposure to radiation or exposure due to radioactive material;
(d) The use of radiation or radioactive material for medical, industrial, veterinary, agricultural, legal or security purposes, including the use of associated equipment, software or devices where such use could affect exposure to radiation;
(e) The use of radiation or radioactive material for education, training or research, including any activities relating to such use that involve or could involve exposure to radiation or exposure due to radioactive material;
(f) The mining and processing of raw materials that involve exposure due to radioactive material;
(g) Any other practice as specified by the regulatory body.

3.2. The requirements for planned exposure situations apply to exposure due to sources within practices[15], as follows:

(a) Facilities that contain radioactive material and facilities that contain radiation generators, including nuclear installations, medical radiation facilities, veterinary radiation facilities, facilities for the management of radioactive waste, installations for the processing of radioactive material, irradiation facilities, and mineral extraction and mineral processing facilities that involve or could involve exposure to radiation or exposure due to radioactive material;

(b) Individual sources of radiation, including sources within the types of facility mentioned in para. 3.2(a), as appropriate, in accordance with the requirements of the regulatory body.

3.3. The requirements for planned exposure situations apply for any occupational exposure, medical exposure or public exposure due to any practice or due to a source within a practice as specified in paras 3.1 and 3.2.

3.4. Exposure due to natural sources is in general considered an existing exposure situation and is subject to the requirements stated in Section 5. However, the relevant requirements in Section 3 for planned exposure situations apply to:

(a) Exposure due to material[16] in any practice specified in para. 3.1 where the activity concentration in the material of any radionuclide in the uranium or thorium decay chains is greater than 1 Bq/g or the activity concentration of ^{40}K is greater than 10 Bq/g;

---

[15] For example, a sterilization gamma irradiation unit is a source for the practice of radiation preservation of food; an X ray unit may be a source for the practice of radiodiagnosis; a nuclear power plant is part of the practice of generating electricity by nuclear fission, and may be regarded as a single source (e.g. with respect to discharges) or as a collection of sources (e.g. for occupational radiation protection purposes). A complex or multiple installation situated at one location or site may, as appropriate, be considered a single source for the purposes of application of these Standards.

(b) Public exposure delivered by discharges or in the management of radioactive waste arising from a practice involving material as specified in para. 3.4(a);
(c) Exposure due to ^{222}Rn and its progeny and ^{220}Rn and its progeny in workplaces in which occupational exposure due to other radionuclides in the uranium or thorium decay chains is controlled as a planned exposure situation;
(d) Exposure due to ^{222}Rn and ^{222}Rn progeny where the annual average activity concentration of ^{222}Rn in air in the workplace remains above the reference level established in accordance with para. 5.27 after the fulfilment of the requirement stated in para. 5.28.

GENERIC REQUIREMENTS

3.5. No person or organization shall adopt, introduce, conduct, discontinue or cease a practice, or shall, as applicable, mine, extract, process, design, manufacture, construct, assemble, install, acquire, import, export, distribute, loan, hire, receive, site, locate, commission, possess, use, operate, maintain, repair, transfer, decommission, disassemble, transport, store or dispose of a source within a practice other than in accordance with the requirements of these Standards.

Requirement 6: Graded approach

The application of the requirements of these Standards in planned exposure situations shall be commensurate with the characteristics of the practice or the source within a practice, and with the magnitude and likelihood of the exposures.

3.6. The application of the requirements of these Standards shall conform to any requirements specified by the regulatory body, in accordance with a graded approach; however, not all the requirements of these Standards are relevant for every practice or source, nor for all the actions specified in para. 3.5.

Requirement 7: Notification and authorization

Any person or organization intending to operate a facility or to conduct an activity shall submit to the regulatory body, as appropriate, a notification or an application for authorization.

Notification

3.7. Any person or organization intending to carry out any of the actions specified in para. 3.5 shall submit a notification to the regulatory body of such an intention[17]. Notification alone is sufficient provided that the exposures expected to be associated with the practice or action are unlikely to exceed a small fraction, as specified by the regulatory body, of the relevant limits, and that the likelihood and magnitude of potential exposures and any other potential detrimental consequences are negligible. Notification is required for consumer products only with respect to manufacture, assembly, maintenance, import, distribution and, in some cases, disposal.

---

[16] A situation of exposure due to radionuclides of natural origin in food, feed, drinking water, agricultural fertilizer and soil amendments, construction material and existing residues in the environment is treated as an existing exposure situation regardless of the activity concentrations of the radionuclides concerned.

[17] With regard to material being transported in accordance with the IAEA Regulations for the Safe Transport of Radioactive Material [12], the requirements of these Standards for notification and authorization are fulfilled by means of compliance with the Regulations.

Authorization: registration or licensing

3.8. Any person or organization intending to carry out any of the actions specified in para. 3.5 shall, unless notification alone is sufficient, apply to the regulatory body for authorization[17], which shall take the form of either registration[18] or licensing.

3.9. Any person or organization applying for authorization:

(a) Shall submit to the regulatory body the relevant information necessary to support the application;
(b) Shall refrain from carrying out any of the actions specified in para. 3.5 until the registration or licence has been granted;
(c) Shall assess the nature, likelihood and magnitude of the expected exposures due to the source and shall take all necessary measures for protection and safety;
(d) Shall, if there is a possibility for an exposure to be greater than a level as specified by the regulatory body, have a safety assessment made and submitted to the regulatory body as part of the application;
(e) Shall, as required by the regulatory body, have an appropriate prospective assessment made for radiological environmental impacts, commensurate with the radiation risks associated with the facility or activity.

Requirement 8: Exemption and clearance

The government or the regulatory body shall determine which practices or sources within practices are to be exempted from some or all of the requirements of these Standards. The regulatory body shall approve which sources, including materials and objects, within notified practices or authorized practices may be cleared from regulatory control.

Exemption

3.10. The government or the regulatory body shall determine which practices or sources within practices are to be exempted from some or all of the requirements of these Standards, including the requirements for notification, registration or licensing, using as the basis for this determination the criteria for exemption specified in Schedule I or any exemption levels specified by the regulatory body on the basis of these criteria.

3.11. Exemption shall not be granted for practices deemed to be not justified.

Clearance

3.12. The regulatory body shall approve which sources, including materials and objects, within notified or authorized practices may be cleared from further regulatory control, using as the basis for such approval the criteria for clearance specified in Schedule I or any clearance levels specified by the regulatory body on the basis of such criteria. By means of this approval the regulatory body shall ensure that sources that have been cleared do not again become subject to the requirements for notification, registration or licensing unless it so specifies.

---

[18] Typical practices that are amenable to registration are those for which: (a) safety can largely be ensured by the design of the facilities and equipment; (b) the operating procedures are simple to follow; (c) the safety training requirements are minimal; and (d) there is a history of few problems with safety in operations. Registration is best suited to those practices for which operations do not vary significantly.

Requirement 9: Responsibilities of registrants and licensees in planned exposure situations

Registrants and licensees shall be responsible for protection and safety in planned exposure situations.

3.13. Registrants and licensees shall bear the responsibility for setting up and implementing the technical and organizational measures that are necessary for protection and safety for the practices and sources for which they are authorized. Registrants and licensees may designate suitably qualified persons to carry out tasks relating to these responsibilities, but they shall retain the prime responsibility for protection and safety. Registrants and licensees shall document the names and responsibilities of persons designated to ensure compliance with the requirements of these Standards.

3.14. Registrants and licensees shall notify the regulatory body of any intention to introduce modifications to any practice or source for which they are authorized, whenever the modifications could have significant implications for protection and safety, and they shall not carry out any such modification unless it is specifically authorized by the regulatory body.

3.15. Registrants and licensees:

(a) Shall establish clear lines of responsibility and accountability for protection and safety for the sources for which they are authorized, and shall establish organizational arrangements for protection and safety;
(b) Shall ensure that any delegation of responsibilities by any other principal party is documented;
(c) Shall, for the sources for which they are authorized and for which a specific safety assessment is required in para. 3.9(d), carry out such an assessment and keep it up to date in accordance with para. 3.35;
(d) Shall, for the sources for which they are authorized and for which the regulatory body requires an assessment to be made of the potential radiological environmental impacts, carry out such an assessment and keep it up to date;
(e) Shall assess the likelihood and magnitude of potential exposures, their likely consequences and the number of persons who may be affected by them;
(f) Shall have in place operating procedures and arrangements for protection and safety that are subject to periodic review and updating under a management system;
(g) Shall establish procedures for reporting on and learning from accidents and other incidents;
(h) Shall establish arrangements for the periodic review of the overall effectiveness of the measures for protection and safety;
(i) Shall ensure that adequate maintenance, testing and servicing are carried out as necessary so that sources remain capable of meeting their design requirements for protection and safety throughout their lifetime;
(j) Shall ensure safe management of and control over all radioactive waste that is generated, and shall dispose of such waste in accordance with the regulatory requirements.

Requirement 10: Justification of practices

The government or the regulatory body shall ensure that only justified practices are authorized.

3.16. The government or the regulatory body, as appropriate, shall ensure that provision[19] is made for the justification of any type of practice[20] and for review of the justification, as necessary, and shall ensure that only justified practices are authorized.

---

[19] Such provision may involve several government entities not necessarily having direct responsibility for protection and safety, such as ministries of health, justice, immigration and security.

[20] This provision for the justification of any type of practice includes practices for which notification alone is

3.17. The following practices are deemed to be not justified:

(a) Practices, except for justified practices involving medical exposure[21], that result in an increase in activity, by the deliberate addition of radioactive substances or by activation[22], in food, feed, beverages, cosmetics or any other commodity or product intended for ingestion, inhalation or percutaneous intake by, or application to, a person;

(b) Practices involving the frivolous use of radiation or radioactive substances in commodities or in products such as toys and personal jewellery or adornments, which result in an increase in activity, by the deliberate addition of radioactive substances or by activation[23];

(c) Human imaging using radiation used as a form of art or for publicity purposes.

3.18. Human imaging using radiation that is performed for occupational, legal or health insurance purposes, and is undertaken without reference to clinical indication, shall normally be deemed to be not justified. If, in exceptional circumstances, the government or the regulatory body decides that the justification of such human imaging for specific practices is to be considered, the requirements of paras 3.61–3.64 and 3.66 shall apply.

3.19. Human imaging using radiation for theft detection purposes shall be deemed to be not justified.

3.20. Human imaging using radiation for the detection of concealed objects for anti-smuggling purposes shall normally be deemed to be not justified. If, in exceptional circumstances, the government or the regulatory body decides that the justification of such human imaging is to be considered, the requirements of paras 3.61–3.67 shall apply.

3.21. Human imaging using radiation for the detection of concealed objects that can be used for criminal acts that pose a national security threat shall be justified only by the government. If the government decides that the justification of such human imaging is to be considered, the requirements of paras. 3.61–3.67 shall apply.

Requirement 11: Optimization of protection and safety

The government or regulatory body shall establish and enforce requirements for the optimization of protection and safety, and registrants and licensees shall ensure that protection and safety is optimized.

3.22. The government or regulatory body:

(a) shall establish and enforce requirements for the optimization of protection and safety;

(b) shall require documentation addressing the optimization of protection and safety;

---

sufficient.

[21] Particular requirements for the justification of medical exposure are specified in paras 3.154–3.160.

[22] This requirement is not intended to prohibit those practices that may involve the short term activation of commodities or products, for which there is no increase in radioactivity in the commodity or product as supplied.

[23] This requirement is not intended to prohibit those practices that may involve the short term activation of commodities or products, for which there is no increase in radioactivity in the commodity or product as supplied.

(c) shall establish or approve constraints[24] on dose and on risk, as appropriate, or shall establish or approve a process for establishing such constraints, to be used in the optimization of protection and safety.

3.23. Registrants and licensees shall ensure that protection and safety is optimized.

3.24. For occupational exposure and public exposure[25], registrants and licensees shall ensure that all relevant factors are taken into account in a coherent way in the optimization of protection and safety to contribute to achieving the following objectives:

(a) To determine measures for protection and safety that are optimized for the prevailing circumstances, with account taken of the available options for protection and safety as well as the nature, likelihood and magnitude of exposures;
(b) To establish criteria, on the basis of the results of the optimization, for the restriction of the likelihood and magnitudes of exposures by means of measures for preventing accidents and for mitigating the consequences of those that do occur.

3.25. For occupational exposure and public exposure, registrants and licensees shall ensure, as appropriate, that relevant constraints are used in the optimization of protection and safety for any particular source within a practice.[24]

Requirement 12: Dose limits

The government or the regulatory body shall establish dose limits for occupational exposure and public exposure, and registrants and licensees shall apply these limits.

3.26. The government or the regulatory body shall establish and the regulatory body shall enforce compliance with the dose limits specified in Schedule III for occupational exposures and public exposures in planned exposure situations.

3.27. The government or the regulatory body shall determine what additional restrictions, if any, are required to be complied with by registrants and licensees to ensure that the dose limits specified in Schedule III are not exceeded owing to possible combinations of doses from exposures due to different authorized practices.

3.28. Registrants and licensees shall ensure that the exposures of individuals due to the practices for which the registrants and licensees are authorized are restricted so that neither the effective dose nor the equivalent dose to tissues or organs exceeds any relevant dose limit specified in Schedule III.[26]

Requirement 13: Safety assessment

The regulatory body shall establish and enforce requirements for safety assessment, and the person or organization responsible for a facility or activity that gives rise to radiation risks shall conduct an appropriate safety assessment of this facility or activity.

---

[24] For occupational exposure, the relevant dose constraint is on individual doses to workers, established and used by registrants and licensees to set the range of options in optimizing protection and safety for the source. For public exposure, the relevant dose constraint is a source related value established or approved by the government or the regulatory body, with account taken of the doses from planned operations of all sources under control. The dose constraint for each particular source is intended, among other things, to ensure that the sum of doses from planned operations for all sources under control remains within the dose limit.

[25] Requirements for the optimization of medical exposure are specified in paras 3.161–3.176.

[26] Dose limits do not apply to medical exposures.

3.29. The regulatory body shall establish requirements for persons or organizations responsible for facilities and activities that give rise to radiation risks to conduct an appropriate safety assessment[27]. Prior to the granting of an authorization, the responsible person or organization shall be required to submit a safety assessment, which shall be reviewed and assessed by the regulatory body.

3.30. The person or organization, as required under para. 3.9(d), or registrants and licensees, as appropriate, shall conduct a safety assessment that is either generic or specific to the practice or source for which they are responsible[28].

3.31. Safety assessments shall be conducted at different stages, including the stages of siting, design, manufacture, construction, assembly, commissioning, operation, maintenance and decommissioning (or closure) of facilities or parts thereof, as appropriate, so as:

(a) To identify the ways in which exposures could be incurred, account being taken of the effects of external events as well as of events directly involving the sources and associated equipment;
(b) To determine the expected magnitudes and likelihood of exposures in normal operation and, to the extent reasonable and practicable, make an assessment of potential exposures;
(c) To assess the adequacy of the provisions for protection and safety.

3.32. The safety assessment shall include, as appropriate, a systematic critical review of:

(a) The operational limits and conditions for the operation of a facility;
(b) The ways in which structures, systems and components, including software, and procedures relating to protection and safety might fail, singly or in combination, or might otherwise give rise to exposures, and the consequences of such events;
(c) The ways in which external factors could affect protection and safety;
(d) The ways in which operating procedures relating to protection and safety might be erroneous, and the consequences of such errors;
(e) The implications for protection and safety of any modifications;
(f) The implications for protection and safety of security measures or of any modifications to security measures;
(g) Any uncertainties or assumptions and their implications for protection and safety.

3.33. The registrant or licensee shall take into account in the safety assessment:

(a) Factors that could precipitate a substantial release of radioactive material, the measures available to prevent or to control such a release, and the maximum activity of radioactive material that, in the event of a major failure of the containment, could be released to the environment;
(b) Factors that could precipitate a smaller but continuing release of radioactive material, and the measures available to detect and to prevent or to control such a release;
(c) Factors that could give rise to unintended operation of any radiation generator or a loss of shielding, and the measures available to detect and to prevent or control such occurrences;
(d) The extent to which the use of redundant and diverse safety features, which are independent of each other so that failure of one does not result in failure of any other, is appropriate to restrict the likelihood and the magnitude of potential exposure.

---

[27] Requirements on safety assessment for facilities and activities are established in Ref. [14].

[28] A generic safety assessment is usually sufficient for types of source with a high degree of uniformity in design. A specific safety assessment is usually required in other cases; however, the specific safety assessment need not include those aspects covered by a generic safety assessment, if a generic safety assessment has been conducted for the source.

3.34. Registrants and licensees shall ensure that the safety assessment is documented and, where appropriate, that it is independently reviewed under the relevant management system.

3.35. Registrants and licensees shall perform additional reviews of the safety assessment as necessary to ensure that the technical specifications or conditions of use continue to be met when:

(a) Significant modifications are envisaged to the facility or to its operating procedures or maintenance procedures;
(b) Significant changes occur on the site that could affect the safety of the facility or of activities on the site;
(c) Information on operating experience, or information about accidents and other incidents that could result in exposures, indicates that the current assessment might be invalid;
(d) Any significant changes in activities are envisaged;
(e) Any relevant changes in guidelines or standards are envisaged or have been made.

3.36. If as a result of a safety assessment, or for any other reason, opportunities to improve protection and safety appear to be available and improvement seems desirable, any consequential modifications shall be made cautiously and only after favourable assessment of all the implications for protection and safety. The implementation of all improvements shall be prioritized so as to optimize protection and safety.

Requirement 14: Monitoring for verification of compliance

Registrants and licensees and employers shall conduct monitoring to verify compliance with the requirements for protection and safety.

3.37. The regulatory body shall establish requirements that monitoring and measurements be performed to verify compliance with the requirements for protection and safety. The regulatory body shall be responsible for the review and approval of monitoring and measurement programmes of registrants and licensees.

3.38. Registrants and licensees and employers shall ensure that:

(a) Monitoring and measurements of parameters are performed as necessary for verification of compliance with the requirements of these Standards;
(b) Suitable equipment is provided and verification procedures are implemented;
(c) Equipment is properly maintained, tested and calibrated at appropriate intervals with reference to standards traceable to national or international standards;
(d) Records are maintained of the results of monitoring and verification of compliance, as required by the regulatory body, including records of the tests and calibrations carried out in accordance with these Standards;
(e) The results of monitoring and verification of compliance are shared with the regulatory body as required.

Requirement 15: Prevention and mitigation of accidents

Registrants and licensees shall apply good engineering practice and shall take all practicable measures to prevent accidents and to mitigate the consequences of those accidents that do occur.

Good engineering practice

3.39. The registrant or licensee, in cooperation with other responsible parties, shall ensure that the siting, location, design, construction, assembly, commissioning, operation, maintenance and decommissioning (or closure) of facilities or parts thereof are based on good engineering practice which shall, as appropriate:

(a) Take account of international and national standards;
(b) Be supported by managerial and organizational features, with the purpose of ensuring protection and safety throughout the lifetime of the facility;
(c) Include adequate safety margins in the design and construction of the facility, and in operations involving the facility, so as to ensure reliable performance in normal operation, and take account of the necessary quality, redundancy and capability for inspection, with emphasis on preventing accidents, mitigating the consequences of those accidents that do occur and restricting any possible future exposures;
(d) Take account of relevant developments concerning technical criteria, as well as the results of any relevant research on protection and safety and feedback of information on lessons learned from experience.

Defence in depth

3.40. Registrants and licensees shall ensure that a multilevel (defence in depth) system of sequential, independent provisions for protection and safety that is commensurate with the likelihood and the magnitude of the potential exposures is applied to sources for which the registrants and licensees are authorized. Registrants and licensees shall ensure that if one level of protection were to fail, the subsequent independent level of protection would be available. Such defence in depth shall be applied for the purposes of:

(a) Preventing accidents;
(b) Mitigating the consequences of any accidents that do occur;
(c) Restoring the sources to safe conditions after any such accidents.

Accident prevention

3.41. Registrants and licensees shall ensure that structures, systems and components, including software, that are related to protection and safety for facilities and activities are designed, constructed, commissioned, operated and maintained so as to prevent accidents as far as reasonably practicable.

3.42. The registrant or licensee for any facility or activity shall make suitable arrangements:

(a) To prevent reasonably foreseeable accidents in the facility or the activity;
(b) To mitigate the consequences of those accidents that do occur;
(c) To provide workers with the information, instruction, training and equipment necessary to restrict potential exposures;
(d) To ensure that there are adequate procedures for the control of the facility and for the management of any reasonably foreseeable accidents;
(e) To ensure that safety significant structures, systems and components, including software, and other equipment can be inspected and tested regularly for any degradation that could lead to abnormal conditions or inadequate performance;
(f) To ensure that maintenance, inspection and testing appropriate to the preservation of the provisions for protection and safety can be carried out without undue occupational exposure;
(g) To provide, wherever appropriate, automatic systems for safely shutting off or reducing the release of radiation from facilities in the event that operating conditions exceed the operating ranges;
(h) To ensure that abnormal operating conditions that could significantly affect protection and safety are detected by systems that respond quickly enough to allow for corrective action to be taken in a timely manner;
(i) To ensure that all relevant safety documentation is available in the appropriate languages.

Emergency preparedness and response

3.43. If the safety assessment indicates that there is a reasonable likelihood of an emergency affecting either workers or members of the public, the registrant or licensee shall prepare an emergency plan for the protection of people and the environment. As part of this emergency plan, the registrant or licensee shall include arrangements for the prompt identification of an emergency, and for determining the appropriate level of the emergency response [15]. In relation to the arrangements for the emergency response at the scene by the registrant or licensee, the emergency plan shall include, in particular:

(a) Provision for individual monitoring and area monitoring and arrangements for medical treatment;
(b) Arrangements for assessing and mitigating any consequences of an emergency.

3.44. Registrants and licensees shall be responsible for the implementation of their emergency plans and shall be prepared to take any necessary action for effective response. To prevent the occurrence of conditions that could lead to a loss of control over a source or to the escalation of such conditions, registrants and licensees shall, as appropriate:

(a) Develop, maintain and implement procedures to provide the means for preventing loss of control over the source and for regaining control over the source as necessary;
(b) Make available equipment, instrumentation and diagnostic aids that may be needed;
(c) Train and periodically retrain personnel in the procedures to be followed and exercise the procedures.

Requirement 16: Investigations and feedback of information on operating experience

Registrants and licensees shall conduct formal investigations of abnormal conditions arising in the operation of facilities or the conduct of activities, and shall disseminate information that is significant for protection and safety.

3.45. Registrants and licensees shall ensure that information on both normal operations and abnormal conditions that are significant for protection and safety is disseminated or made available, as appropriate, to the regulatory body and relevant parties, as specified by the regulatory body. This information would include, for example, details of doses associated with given activities, data on maintenance, descriptions of events and information on corrective actions, and information on operating experience from other relevant facilities and activities.

3.46. Registrants and licensees shall conduct an investigation as specified by the regulatory body in the event that:

(a) A quantity or operating parameter relating to protection and safety exceeds an investigation level or is outside the stipulated range of operating conditions; or
(b) Any equipment failure, accident, error, mishap or other unusual event or condition occurs that has the potential for causing a quantity to exceed any relevant limit or operating restriction.

3.47. The registrant or licensee shall conduct an investigation as soon as possible after an event and shall prepare a written record of its causes, or suspected causes, including a verification or determination of any doses received or committed and recommendations for preventing the recurrence of the event and the occurrence of similar events.

3.48. The registrant or licensee shall communicate to the regulatory body and to any other relevant parties, as appropriate, a written report of any formal investigation relating to events prescribed by the regulatory body, including exposures greater than a dose limit. The registrant or licensee also shall immediately report any event in which a dose limit is exceeded.

Requirement 17: Radiation generators and radioactive sources

Registrants and licensees shall ensure the safety of radiation generators and radioactive sources.

3.49. Registrants and licensees who are manufacturers or other suppliers of radiation generators or radiation sources shall ensure that the following responsibilities are discharged, as applicable:

(a) Supplying a well designed, well manufactured and well constructed radiation generator or radioactive source and device in which the radiation generator or radioactive source is used that:
   (i) Provides for protection and safety in accordance with the requirements of these Standards;
   (ii) Meets engineering, performance and functional specifications;
   (iii) Meets quality standards commensurate with the significance for protection and safety of systems and components, including software;
   (iv) Provides clear displays, gauges and instructions on operating consoles in the appropriate language.
(b) Ensuring that radiation generators and radioactive sources are tested to demonstrate compliance with the relevant specifications;
(c) Making information available, in the appropriate language, on the proper installation and use of the radiation generator or radioactive source and its associated radiation risks, including performance specifications, instructions for operating and maintenance, and instructions for protection and safety;
(d) Ensuring that the protection provided by shielding and other protective devices is optimized.

3.50. Where applicable, registrants and licensees shall make suitable arrangements with suppliers of radiation generators and radioactive sources, the regulatory body and relevant parties for the purposes of:

(a) obtaining information on conditions of use and operating experience that may be important for protection and safety;
(b) providing feedback and information that may have implications for protection and safety for other users, or that may have implications for the possibility for improvements in protection and safety for radiation generators and radioactive sources.

3.51. When choosing a location to use or to store a radiation generator or radioactive source, registrants and licensees shall take into account:

(a) Factors that could affect the safe management of and control over the radiation generator or radioactive source;
(b) Factors that could affect occupational exposure and public exposure due to the radiation generator or radioactive source;
(c) The feasibility of taking the foregoing factors into account in engineering design.

3.52. In selecting a site for a facility that will contain a large amount of radioactive material and will have the potential for the release of significant amounts of radioactive material, registrants and licensees shall take into account features that might affect protection and safety, features that might affect the integrity or functioning of the facility, and the feasibility of carrying out off-site protective actions if they become necessary.

3.53. Registrants and licensees shall keep radiation generators and radioactive sources under control so as to prevent loss or damage and to prevent any unauthorized person from carrying out any of the activities specified in para. 3.5, by ensuring that:

(a) Control over a radiation generator or radioactive source is relinquished only in compliance with all relevant requirements specified in the registration or licence;
(b) The regulatory body is promptly notified of information regarding a radiation generator or radioactive source that is lost, missing or not under control;
(c) A radiation generator or radioactive source is transferred only if the receiver possesses the necessary authorization;
(d) An inventory, as required in para. 3.54, of radiation generators or radioactive sources is conducted periodically to confirm that they are in their assigned locations and are under control.

3.54. Registrants and licensees shall maintain an inventory that includes records of:

(a) The location and description of each radiation generator or radioactive source for which they are responsible;
(b) The activity and form of each radioactive source for which they are responsible.

3.55. Registrants and licensees shall provide the regulatory body as required with appropriate information from their inventory records of radiation generators and radioactive sources.

3.56. Registrants and licensees shall ensure that sealed sources are categorized in accordance with the categorization scheme set out in Schedule II, and in accordance with the requirements of the regulatory body.

3.57. The manufacturer of a radioactive source or a device containing a radioactive source shall ensure that, where practicable, the source itself and its container are marked with the symbol recommended by the International Organization for Standardization (ISO) [16][29].

3.58. Registrants and licensees, in cooperation with manufacturers, shall ensure that, where practicable, sealed sources are identifiable and traceable.

3.59. Registrants and licensees shall ensure that when radioactive sources are not in use they are stored in an appropriate manner for protection and safety.

3.60. Registrants and licensees shall ensure that arrangements are made promptly for the safe management of and control over radiation generators and radioactive sources, including appropriate financial provision, once it has been decided to take them out of use.

Requirement 18: Human imaging using radiation for purposes other than medical diagnosis, medical treatment or biomedical research

The government shall ensure that the use of ionizing radiation for human imaging for purposes other than medical diagnosis, medical treatment or biomedical research is subject to the system of protection and safety.

3.61. The government, if so decided in accordance with paras 3.18, 3.20 and 3.21, shall ensure that the requirements of para. 3.16 for the justification of practices are applied to any type of human imaging procedure in which radiation is used for purposes other than for medical diagnosis or medical

---

[29] For Category 1, 2 and 3 sealed sources as defined in Schedule II, the manufacturer may consider the placement near the source, preferably on the shield or near the point of access to the source, of the supplementary symbol specified in Ref. [17]. The supplementary symbol is not placed on the external surfaces of transport packages, freight containers or conveyances or on building access doors.

treatment or as part of a programme of biomedical research. The justification process shall include the consideration of:

(a) The benefits and detriments of implementing the type of human imaging procedure;
(b) The benefits and detriments of not implementing the type of human imaging procedure;
(c) Any legal or ethical issues associated with the introduction of the type of human imaging procedure;
(d) The effectiveness and suitability of the type of human imaging procedure, including the appropriateness of the radiation equipment for the intended use;
(e) The availability of sufficient resources to conduct the human imaging procedure safely throughout the intended period of the practice.

3.62. If it has been determined through the process specified in para. 3.61 that a particular practice of human imaging using radiation is justified, then, such a practice shall be subject to regulatory control.

3.63. The regulatory body, in cooperation with other relevant authorities, agencies and professional bodies, as appropriate, shall establish the requirements for regulatory control of the practice, and for review of the justification.

3.64. For human imaging using radiation conducted by medical personnel using medical radiological equipment, which exposes humans to radiation for employment related, legal or health insurance[30] purposes without reference to clinical indications:

(a) The government shall ensure, on the basis of consultation between relevant authorities, professional bodies and the regulatory body, that dose constraints are established for such human imaging;
(b) The registrant or licensee shall ensure that the appropriate optimization requirements for medical exposure in paras 3.161–3.176 are applied, with dose constraints as required in (a) above used instead of diagnostic reference levels.

3.65. Procedures with inspection imaging devices in which radiation is used to expose persons for the purpose of detection of concealed weapons, contraband or other objects on or within the body shall be considered to give rise to public exposure. Registrants and licensees shall apply the requirements for public exposure in planned exposure situations. In particular, registrants and licensees shall ensure that optimization of protection and safety is subject to any dose constraints for public exposure set by the government or the regulatory body.

3.66. Registrants and licensees shall ensure that all persons who are to undergo procedures with inspection imaging devices in which ionizing radiation is used are informed of the possibility of requesting the use of an alternative inspection technique that does not use ionizing radiation, where available.

3.67. The registrant or licensee shall ensure that any inspection imaging device used for the detection of concealed objects on or within the body, whether it is manufactured in or imported into the State in which it is used, conforms to applicable standards of the International Electrotechnical Commission or the International Organization for Standardization or to equivalent national standards.

---

[30] Such purposes include assessment of fitness for employment (prior to employment or periodically during employment), assessment of physiological suitability for a career or a sport, assessment of athletes before a selection or transfer, determination of age for legal purposes, obtaining of evidence for legal purposes, detection of drugs concealed within the body, immigration or emigration requirements, pre-insurance checks and obtaining evidence for the purposes of a compensation claim.

OCCUPATIONAL EXPOSURE

Scope

3.68. The requirements in respect of occupational exposure in planned exposure situations (paras 3.69–3.116) apply to occupational exposure due to a practice or a source within a practice, as stated in paras 3.1–3.3; and to occupational exposure as required in Section 4 for emergency exposure situations and as required in Section 5 for existing exposure situations. For exposure due to natural sources, these requirements for occupational exposure in planned exposure situations apply, as appropriate, only to the exposure situations specified in para. 3.4 (a), (c) and (d).

Requirement 19: Responsibilities of the regulatory body specific to occupational exposure

The government or regulatory body shall establish and enforce requirements to ensure that protection and safety is optimized, and the regulatory body shall enforce compliance with dose limits for occupational exposure.

3.69. The government or regulatory body shall establish the responsibilities of employers, registrants and licensees with regard to application of the requirements for occupational exposure in planned exposure situations.

3.70. The government or regulatory body shall establish and enforce requirements to ensure that protection and safety is optimized for occupational exposure.

3.71. The government or regulatory body shall establish and the regulatory body shall enforce compliance with the dose limits specified in Schedule III for occupational exposure.

3.72. Before authorization of a new or modified practice, the regulatory body shall require, as appropriate, and review supporting documents from the responsible parties that state:

(a) design criteria and design features relating to the exposure and potential exposure of workers in all operational states and accident conditions;
(b) design criteria and design features of the appropriate systems and programmes for monitoring of workers for occupational exposure in all operational states and accident conditions.

Requirement 20: Requirements for monitoring and recording of occupational exposure

The regulatory body shall establish and enforce requirements for the monitoring and recording of occupational exposures in planned exposure situations.

3.73. The regulatory body shall be responsible, as appropriate, for:

(a) Establishment and enforcement of requirements for the monitoring, recording and control of occupational exposures in planned exposure situations in accordance with the requirements of these Standards;
(b) Review of monitoring programmes of registrants and licensees, which shall be adequate to ensure that the requirements with regard to occupational exposure in planned exposure situations are met;
(c) Authorization or approval of service providers for individual monitoring and calibration services;
(d) Review of periodic reports on occupational exposure (including results of monitoring programmes and dose assessments) submitted by employers, registrants and licensees;
(e) Provision for maintaining exposure records and results of the assessment of doses from occupational exposure;

(f) Verification of compliance of an authorized practice with the requirements on the control of occupational exposure.

Requirement 21: Responsibilities of employers, registrants and licensees for the protection of workers

Employers, registrants and licensees shall be responsible for the protection of workers against occupational exposure. Employers, registrants and licensees shall ensure that protection and safety is optimized and that the dose limits for occupational exposure are not exceeded.

3.74. For workers who are engaged in activities in which they are or could be subject to occupational exposure in planned exposure situations, employers, registrants and licensees shall be responsible for:

(a) Protection of workers against occupational exposure;
(b) Compliance with other relevant requirements of these Standards.

3.75. Employers who are also registrants or licensees shall have the responsibilities of both employers and registrants or licensees.

3.76. Employers, registrants and licensees shall ensure, for all workers engaged in activities in which they are or could be subject to occupational exposure, that:

(a) Occupational exposure is controlled so that the relevant dose limits for occupational exposure specified in Schedule III are not exceeded;
(b) Protection and safety is optimized in accordance with the requirements of these Standards;
(c) Decisions with regard to measures for protection and safety are recorded and made available to relevant parties, through their representatives where appropriate, as specified by the regulatory body;
(d) Policies, procedures and organizational arrangements for protection and safety are established for implementing the relevant requirements of these Standards, with priority given to design measures and technical measures for controlling occupational exposure;
(e) Suitable and adequate facilities, equipment and services for protection and safety are provided, the type and extent of which are commensurate with the expected likelihood and magnitude of the occupational exposure;
(f) Necessary health surveillance and health services for workers are provided;
(g) Appropriate monitoring equipment and personal protective equipment are provided and arrangements are made for its proper use, calibration, testing and maintenance;
(h) Suitable and adequate human resources and appropriate training in protection and safety are provided, as well as periodic retraining as required to ensure the necessary level of competence;
(i) Adequate records are maintained in accordance with the requirements of these Standards;
(j) Arrangements are made to facilitate consultation of and cooperation with workers, through their representatives where appropriate, with regard to protection and safety on all measures necessary to achieve the effective application of these Standards;
(k) Necessary conditions for promoting a safety culture are provided.

3.77. Employers, registrants and licensees shall:

(a) Involve workers, through their representatives where appropriate, in optimization of protection and safety;
(b) Establish and use, as appropriate, constraints as part of optimization of protection and safety.

3.78. Employers, registrants and licensees shall ensure that workers exposed to radiation from sources within a practice that are not required by or directly related to their work have the same level of protection against such exposure as members of the public.

3.79. Employers, registrants and licensees shall take such administrative actions as are necessary to ensure that workers are informed that ensuring protection and safety is an integral part of a general occupational health and safety programme in which they have specific obligations and responsibilities for their own protection and the protection of others against radiation exposure and for the safety of sources.

3.80. Employers, registrants and licensees shall record any report received from a worker that identifies circumstances that could affect compliance with the requirements of these Standards, and shall take appropriate action.

3.81. Nothing in these Standards shall be construed as relieving employers from complying with applicable national and local laws and regulations governing hazards in the workplace.

3.82. Employers, registrants and licensees shall facilitate compliance by workers with the requirements of these Standards.

Requirement 22: Compliance by workers

Workers shall fulfil their obligations and carry out their duties for protection and safety.

3.83. Workers:

(a) Shall follow any applicable rules and procedures for protection and safety as specified by the employer, registrant or licensee;
(b) Shall use properly the monitoring equipment and personal protective equipment provided;
(c) Shall cooperate with the employer, registrant or licensee with regard to protection and safety, and programmes for workers' health surveillance and programmes for dose assessment;
(d) Shall provide to the employer, registrant or licensee such information on their past and present work that is relevant for ensuring effective and comprehensive protection and safety for themselves and others;
(e) Shall abstain from any wilful action that could put themselves or others in situations that would not be in accordance with the requirements of these Standards;
(f) Shall accept such information, instruction and training in protection and safety as will enable them to conduct their work in accordance with the requirements of these Standards.

3.84. A worker who identifies circumstances that could adversely affect protection and safety shall report such circumstances to the employer, registrant or licensee as soon as possible.

Requirement 23: Cooperation between employers and registrants and licensees

Employers and registrants and licensees shall cooperate to the extent necessary for compliance by all responsible parties with the requirements for protection and safety.

3.85. If workers are engaged in work that involves or that could involve a source that is not under the control of their employer, the registrant or licensee responsible for the source and the employer shall cooperate to the extent necessary for compliance by both parties with the requirements of these Standards.

3.86. Cooperation between the employer and the registrant or licensee shall include, where appropriate:

(a) The development and use of specific restrictions on exposure and other means of ensuring that the measures for protection and safety for workers who are engaged in work that involves or could involve a source that is not under the control of their employer are at least as good as those for employees of the registrant or licensee;

(b) Specific assessments of the doses received by workers as specified in (a);
(c) A clear allocation and documentation of the responsibilities of the employer and those of the registrant or licensee for protection and safety.

3.87. As part of the cooperation between parties, the registrant or licensee responsible for the source or for the exposure shall, as appropriate:

(a) Obtain from the employers, including self-employed individuals, the previous occupational exposure history of workers as specified in para 3.86, and any other necessary information;
(b) Provide appropriate information to the employer, including any available information relevant for compliance with the requirements of these Standards that the employer requests;
(c) Provide both the worker and the employer with the relevant exposure records.

Requirement 24: Arrangements under the radiation protection programme

Employers, registrants and licensees shall establish and maintain organizational, procedural and technical arrangements for the designation of controlled areas and supervised areas, for local rules and for monitoring of the workplace, in a radiation protection programme for occupational exposure.

Classification of areas: controlled areas

3.88. Registrants and licensees shall designate as a controlled area any area[31] in which specific measures for protection and safety are or could be required for:

(a) Controlling exposures or preventing the spread of contamination in normal operation;
(b) Preventing or limiting the likelihood and magnitude of exposures in anticipated operational occurrences and accident conditions.

3.89. In defining the boundaries of any controlled area, registrants and licensees shall take account of the magnitudes of the exposures expected in normal operation, the likelihood and magnitude of exposures in anticipated operational occurrences and in accident conditions, and the type and extent of the procedures required for protection and safety.

3.90. Registrants and licensees:

(a) Shall delineate controlled areas by physical means or, where this is not reasonably practicable, by some other suitable means;
(b) Shall, where a source is only intermittently brought into operation or energized, or is moved from place to place, delineate an appropriate controlled area by means that are appropriate under the prevailing circumstances and shall specify exposure times;
(c) Shall display the symbol recommended by the International Organization for Standardization (ISO) [16] and shall display instructions at access points to and at appropriate locations within controlled areas;
(d) Shall establish measures for protection and safety, including, as appropriate, physical measures to control the spread of contamination and local rules and procedures for controlled areas;
(e) Shall restrict access to controlled areas by means of administrative procedures such as the use of work permits, and by physical barriers, which could include locks or interlocks, the degree of restriction being commensurate with the likelihood and magnitude of exposures;
(f) Shall provide, as appropriate, at entrances to controlled areas:
    (i) Personal protective equipment;

---

[31]The transport of radioactive material is regulated in accordance with the IAEA Regulations for the Safe Transport of Radioactive Material [12].

(ii) Equipment for individual monitoring and workplace monitoring;
(iii) Suitable storage for personal clothing;

(g) Shall provide, as appropriate, at exits from controlled areas:
(i) Equipment for monitoring for contamination of skin and clothing;
(ii) Equipment for monitoring for contamination of any objects or material being removed from the area;
(iii) Washing or showering facilities and other personal decontamination facilities;
(iv) Suitable storage for contaminated personal protective equipment;

(h) Shall periodically review conditions to assess whether there is any need to modify the measures for protection and safety or the boundaries of controlled areas;

(i) Shall provide appropriate information, instruction and training for persons working in controlled areas.

Classification of areas: supervised areas

3.91. Registrants and licensees shall designate as a supervised area any area not already designated as a controlled area but for which occupational exposure conditions need to be kept under review, even though specific measures for protection and safety are not normally needed.

3.92. Registrants and licensees, taking into account the nature, likelihood and magnitude of exposures or contamination in the supervised areas:

(a) Shall delineate the supervised areas by appropriate means;
(b) Shall display approved signs, as appropriate, at access points to supervised areas;
(c) Shall periodically review conditions to assess whether there is any need for further measures for protection and safety or any need for changes to the boundaries of supervised areas.

Local rules and procedures and personal protective equipment

3.93. Employers, registrants and licensees shall minimize the need to rely on administrative controls and personal protective equipment for protection and safety by providing well engineered controls and satisfactory working conditions, in accordance with the following hierarchy of preventive measures:

(1) Engineered controls;
(2) Administrative controls;
(3) Personal protective equipment.

3.94. Employers, registrants and licensees, in consultation with workers or through their representatives:

(a) Shall establish in writing local rules and procedures that are necessary for protection and safety for workers and other persons;
(b) Shall include in the local rules and procedures any relevant investigation level or authorized level, and the procedures to be followed in the event that any such level is exceeded;
(c) Shall make the local rules and procedures and the measures for protection and safety known to those workers to whom they apply and to other persons who may be affected by them;
(d) Shall ensure that any work in which workers are or could be subject to occupational exposure is adequately supervised and shall take all reasonable steps to ensure that the rules, procedures, and measures for protection and safety are observed;
(e) Shall designate, as appropriate, a radiation protection officer in accordance with criteria established by the regulatory body.

3.95. Employers, registrants and licensees shall ensure that:

(a) Workers are provided with suitable and adequate personal protective equipment that meets relevant standards or specifications, including as appropriate:
   (i) Protective clothing;
   (ii) Respiratory protective equipment the characteristics of which are made known to the users;
   (iii) Protective aprons, protective gloves and organ shields;
(b) Where appropriate, workers receive adequate instruction in the proper use of respiratory protective equipment, including testing for good fit;
(c) Tasks requiring the use of certain personal protective equipment are assigned only to workers who on the basis of medical advice are capable of safely sustaining the extra effort necessary;
(d) All personal protective equipment, including equipment for use in an emergency, is maintained in proper condition and, if appropriate, is tested at regular intervals;
(e) If the use of personal protective equipment is considered for any given task, account is taken of any additional exposure that could result owing to the additional time taken or the inconvenience, and of any non-radiological risks that might be associated with using personal protective equipment while performing the task.

Monitoring of the workplace

3.96. Registrants and licensees, in cooperation with employers where appropriate, shall establish, maintain and keep under review a programme for workplace monitoring under the supervision of a radiation protection officer or qualified expert.

3.97. The type and frequency of workplace monitoring shall:

(a) Be sufficient to enable:
   (i) Evaluation of the radiological conditions in all workplaces;
   (ii) Assessment of exposures in controlled areas and supervised areas;
   (iii) Review of the classification of controlled areas and supervised areas;
(b) Be based on dose rate, activity concentration in air and surface contamination, and their expected fluctuations, and on the likelihood and magnitude of exposures in anticipated operational occurrences and accident conditions.

3.98. Registrants and licensees, in cooperation with employers where appropriate, shall maintain records of the findings of the workplace monitoring programme. The findings of the workplace monitoring programme shall be made available to workers, where appropriate through their representatives.

Requirement 25: Assessment of occupational exposure and workers' health surveillance

Employers, registrants and licensees shall be responsible for making arrangements for assessment and recording of the occupational exposure and for workers' health surveillance.

Occupational exposure assessment

3.99. Employers, as well as self-employed persons, and registrants and licensees shall be responsible for making arrangements for assessment of the occupational exposure of workers, on the basis of individual monitoring where appropriate, and shall ensure that arrangements are made with authorized or approved dosimetry service providers that operate under a quality management system.

3.100. For any worker who usually works in a controlled area, or who occasionally works in a controlled area and may receive a significant dose from occupational exposure, individual monitoring shall be undertaken where appropriate, adequate and feasible. In cases where individual monitoring of the worker is inappropriate, inadequate or not feasible, the occupational exposure shall be assessed on the basis of the results of workplace monitoring and information on the locations and durations of exposure of the worker[32].

3.101. For any worker who regularly works in a supervised area or who enters a controlled area only occasionally, the occupational exposure shall be assessed on the basis of the results of workplace monitoring or individual monitoring, as appropriate.

3.102. Employers shall ensure that workers who could be subject to exposure due to contamination are identified, including workers who use respiratory protective equipment. Employers shall arrange for appropriate monitoring to the extent necessary to demonstrate the effectiveness of the measures for protection and safety and to assess intakes of radionuclides and the committed effective doses.

Records of occupational exposure

3.103. Employers, registrants and licensees shall maintain records of occupational exposure[33] for every worker for whom assessment of occupational exposure is required in paras 3.99–3.102.

3.104. Records of occupational exposure for each worker shall be maintained during and after the worker's working life, at least until the former worker attains or would have attained the age of 75 years, and for not less than 30 years after cessation of the work in which the worker was subject to occupational exposure.

3.105. Records of occupational exposure shall include:

(a) Information on the general nature of the work in which the worker was subject to occupational exposure;
(b) Information on dose assessments, exposures and intakes at or above the relevant recording levels and the data upon which the dose assessments were based;
(c) When a worker is or has been exposed while in the employ of more than one employer, information on the dates of employment with each employer and on the doses, exposures and intakes in each such employment;
(d) Records of any assessments of doses, exposures and intakes due to actions taken in an emergency or due to accidents or other incidents, which shall be distinguished from assessments of doses, exposures and intakes due to normal conditions of work and which shall include references to reports of any relevant investigations.

3.106. Employers, registrants and licensees:

(a) Shall provide workers with access to records of their own occupational exposure;
(b) Shall provide the supervisor of the programme for workers' health surveillance, the regulatory body and the relevant employer with access to workers' records of occupational exposure;
(c) Shall facilitate the provision of copies of workers' exposure records to new employers when workers change employment;
(d) Shall make arrangements for the retention of exposure records for former workers by the employer, registrant or licensee, as appropriate;
(e) Shall, in complying with (a)–(d) above, give due care and attention to maintaining the confidentiality of records.

3.107. If employers, registrants and licensees cease to conduct activities in which workers are subject to occupational exposure, they shall make arrangements for the retention of workers' records

---

[32] The distinction between types of workers in paras 3.100 and 3.101 for the purposes of monitoring has similarities to the distinction between Category A and Category B workers in European Union legislation [18].

[33] Records of occupational exposure are also referred to as 'exposure records' or 'dose records'.

of occupational exposure by the regulatory body or a State registry, or by a relevant employer, registrant or licensee, as appropriate.

Workers' health surveillance

3.108. Programmes for workers' health surveillance as required in para. 3.76(f):

(a) Shall be based on the general principles of occupational health [19];
(b) Shall be designed to assess the initial fitness and continuing fitness of workers for their intended tasks.

3.109. If one or more workers are to be engaged in work in which they are or could be exposed to radiation from a source that is not under the control of their employer, the registrant or licensee responsible for the source shall, as a precondition for the engagement of such workers, make any special arrangements for workers' health surveillance with the employer that are needed to comply with the rules established by the regulatory body or other relevant authority.

Requirement 26: Information, instruction and training

Employers, registrants and licensees shall provide workers with adequate information, instruction and training for protection and safety.

3.110. Employers, in cooperation with registrants and licensees:

(a) Shall provide all workers with adequate information on health risks due to their occupational exposure in normal operation, anticipated operational occurrences and accident conditions, adequate instruction and training and periodic retraining in protection and safety, and adequate information on the significance of their actions for protection and safety;
(b) Shall provide those workers who could be involved in or affected by the response to an emergency with appropriate information, and adequate instruction and training and periodic retraining, for protection and safety;
(c) Shall maintain records of the training provided to individual workers.

Requirement 27: Conditions of service

Employers, registrants and licensees shall not offer benefits as substitutes for measures for protection and safety.

3.111. The conditions of service of workers shall be independent of whether they are or could be subject to occupational exposure. Special compensatory arrangements, or preferential consideration with respect to salary, special insurance coverage, working hours, length of vacation, additional holidays or retirement benefits, shall neither be granted nor be used as substitutes for measures for protection and safety in accordance with the requirements of these Standards.

3.112. Employers shall make all reasonable efforts to provide workers with suitable alternative employment in circumstances for which it has been determined, either by the regulatory body or in the framework of the programme for workers' health surveillance in accordance with the requirements of these Standards, that workers, for health reasons, may no longer continue in employment in which they are or could be subject to occupational exposure.

Requirement 28: Special arrangements

Employers, registrants and licensees shall make special arrangements for female workers, as necessary, for protection of the embryo or fetus and of breast-fed infants. Employers, registrants and licensees shall make special arrangements for protection and safety for persons under 18 years of age who are undergoing training.

3.113. Employers, in cooperation with registrants and licensees, shall provide female workers who are liable to enter controlled areas or supervised areas or who may undertake emergency duties with appropriate information on:

(a) The risk to the embryo or fetus due to exposure of a pregnant woman;
(b) The importance for a female worker of notifying her employer as soon as possible if she suspects that she is pregnant[34] or if she is breast-feeding;
(c) The risk of health effects for a breast-fed infant due to ingestion of radioactive substances.

3.114. Notification of the employer by a female worker if she suspects that she is pregnant or if she is breast-feeding shall not be considered a reason to exclude a female worker from work. The employer of a female worker, who has been notified of her suspected pregnancy or that she is breast-feeding, shall adapt the working conditions in respect of occupational exposure so as to ensure that the embryo or fetus or the infant is afforded the same broad level of protection as is required for members of the public.

3.115. Employers, registrants and licensees shall ensure that no person under the age of 16 years is or could be subject to occupational exposure.

3.116. Employers, registrants and licensees shall ensure that persons under the age of 18 years are allowed access to a controlled area only under supervision and only for the purpose of training for employment in which they are or could be subject to occupational exposure or for the purpose of studies in which sources are used.

PUBLIC EXPOSURE

Scope

3.117. The requirements in respect of public exposure in planned exposure situations (paras 3.117–3.143) apply to public exposure due to a practice or a source within a practice, as referred to in paras 3.1–3.3. For exposure due to natural sources, such requirements apply only to the types of public exposure specified in para. 3.4 (a) and (b).

Requirement 29: Responsibilities of the government and the regulatory body specific to public exposure

The government or the regulatory body shall establish the responsibilities of relevant parties that are specific to public exposure, shall establish and enforce requirements for optimization, and shall establish, and the regulatory body shall enforce compliance with, dose limits for public exposure.

3.118. The government or the regulatory body shall establish the responsibilities of registrants, licensees, suppliers, and providers of consumer products[35] in relation to the application of requirements for public exposure in planned exposure situations.

---

[34] Notification of an employer of a suspected pregnancy or of breast-feeding cannot be a requirement on a female worker in these Standards. However, it is important that all female workers understand the importance of making such notifications so that their working conditions may be modified accordingly.

[35] ‘Providers of consumer products’ include the designers, manufacturers, producers, constructors, assemblers, installers, distributors, sellers and importers of consumer products.

3.119. The government or the regulatory body shall establish and enforce requirements for the optimization of protection and safety for situations in which individuals are or could be subject to public exposure.

3.120. The government or the regulatory body shall establish or approve constraints on dose and on risk to be used in the optimization of protection and safety for members of the public. When establishing or approving constraints in respect of a source within a practice, the government or the regulatory body shall take into account, as appropriate:

(a) The characteristics of the source and of the practice that are of relevance for public exposure;
(b) Good practice in the operation of similar sources;
(c) Dose contributions from other authorized practices or from possible future authorized practices[36], estimated at the design and planning stage, so that the total dose to members of the public is not expected to exceed the dose limit at any time after the start of operation of the source;
(d) The views of interested parties.

3.121. The government or the regulatory body shall establish, and the regulatory body shall enforce compliance with, the dose limits specified in Schedule III for public exposure.

3.122. Before authorization of a new or modified practice, the regulatory body shall require the submission of, and shall review, the safety assessments (paras 3.29–3.36) and other design related documents from the responsible parties that address the optimization of protection and safety, the design criteria and the design features relating to the assessment of exposure and potential exposure of members of the public.

3.123. The regulatory body shall establish or approve operational limits and conditions relating to public exposure, including authorized limits for discharges. These operational limits and conditions:

(a) Shall be used by registrants and licensees as the criteria for demonstration of compliance after the commencement of operation of a source;
(b) Shall correspond to doses below the dose limits with account taken of the results of optimization of protection and safety;
(c) Shall reflect good practice in the operation of similar facilities or activities;
(d) Shall allow for operational flexibility;
(e) Shall take into account the results of the assessment of the potential radiological environmental impacts undertaken in accordance with national requirements.

3.124. When a source within a practice could cause public exposure outside the territory or other area under the jurisdiction or control of the State in which the source is located, the government or the regulatory body:

(a) Shall ensure that the assessment of the radiological impacts includes those impacts outside the territory or other area under the jurisdiction or control of the State;
(b) Shall, to the extent possible, establish requirements for the control of discharges;
(c) Shall arrange with the affected State the means for the exchange of information and consultations, as appropriate.

---

[36] Dose contributions from possible future authorized practices have to be anticipated in an assessment made on the basis of realistic assumptions.

Requirement 30: Responsibilities of relevant parties specific to public exposure

Relevant parties shall apply the system of protection and safety to protect members of the public against exposure.

General considerations

3.125. Registrants and licensees in cooperation with suppliers and with providers of consumer products shall apply the requirements of these Standards and shall verify and demonstrate compliance with them, as specified by the regulatory body, in relation to any public exposure delivered by a source for which they have responsibility.

3.126. Registrants and licensees in cooperation with suppliers, in applying the principle of optimization of protection and safety in the design, planning, operation and decommissioning of a source (or for closure and the post-closure period for waste disposal facilities), shall take into account:

(a) Possible changes in any conditions that could affect exposure of members of the public, such as changes in the characteristics and use of the source, changes in environmental dispersion conditions, changes in exposure pathways or changes in values of parameters used for the determination of the representative person;
(b) Good practice in the operation of similar sources or the conduct of similar practices;
(c) Possible buildup and accumulation in the environment of radioactive substances from discharges during the lifetime of the source;
(d) Uncertainties in the assessment of doses, especially uncertainties in contributions to doses if the source and the representative person are separated in space or in time.

3.127. Registrants and licensees, for sources under their responsibility, shall establish, implement and maintain:

(a) Policies, procedures and organizational arrangements for protection and safety in relation to public exposure, in accordance with the requirements of these Standards;
(b) Measures for ensuring:
    (i) optimization of protection and safety;
    (ii) limitation of exposure of members of the public from such sources, in accordance with the authorization;
(c) Measures for ensuring the safety of such sources;
(d) Provision for suitable and adequate resources (including facilities, equipment and services) for the protection and safety of members of the public, commensurate with the magnitude and likelihood of exposures;
(e) Programmes for appropriate training of personnel having functions relevant to protection and safety of members of the public, as well as periodic retraining as required, to ensure the necessary level of competence;
(f) Provision for appropriate monitoring equipment, surveillance programmes and methods for assessing public exposure;
(g) Adequate records of surveillance and monitoring;
(h) Emergency plans, emergency procedures and emergency response arrangements, in accordance with the nature and magnitude of the radiation risks associated with the sources.

Visitors

3.128. Registrants and licensees, in cooperation with employers where appropriate:

(a) Shall apply the relevant requirements of these Standards in respect of public exposure for visitors to a controlled area or a supervised area;

(b) Shall ensure that visitors are accompanied in any controlled area by a person who knows the measures for protection and safety for the controlled area;
(c) Shall provide adequate information and instructions to visitors before they enter a controlled area or a supervised area so as to provide for protection and safety for visitors and other individuals who could be affected by their actions;
(d) Shall ensure that adequate control is maintained over the entry of visitors to a controlled area or a supervised area, including the use of signs for such areas.

External exposure and contamination in areas accessible to members of the public

3.129. Registrants and licensees shall ensure that if a source can give rise to external exposure of members of the public:

(a) The floor plans and arrangements of equipment for all new installations utilizing such sources, as well as all significant modifications to existing installations, are subject, as appropriate, to review and approval by the regulatory body prior to commissioning;
(b) Shielding and other protective measures, including access control, are provided as appropriate for restricting public exposure, in particular at open sites such as for some applications of industrial radiography.

3.130. Registrants and licensees shall ensure, as appropriate, that:

(a) Specific provisions for confinement are established for the design and operation of a source that could cause the spread of contamination in areas that are accessible to members of the public;
(b) Protective measures are implemented for restricting public exposure due to contamination in areas within a facility that are accessible to members of the public.

Requirement 31: Radioactive waste and discharges

Relevant parties shall ensure that radioactive waste and discharges of radioactive material to the environment are managed in accordance with the authorization.

Radioactive waste

3.131. Registrants and licensees, in cooperation with suppliers, as appropriate:

(a) Shall ensure that any radioactive waste generated is kept to the minimum practicable in terms of both activity and volume;
(b) Shall ensure that radioactive waste is managed in accordance with the requirements of these Standards and the requirements of other applicable IAEA standards, and in accordance with the relevant authorization;
(c) Shall ensure that there is separate processing of radioactive waste of different types, where warranted by differences in factors such as radionuclide content, half-life, activity concentration, volume, and physical and chemical properties, taking into account the available options for waste storage and disposal, without precluding the mixing of waste for purposes of protection and safety;
(d) Shall ensure that activities for the predisposal management of and for the disposal of radioactive waste are conducted in accordance with the requirements of applicable IAEA standards[37], and in accordance with the authorization;
(e) Shall maintain an inventory of all radioactive waste that is generated, stored, transferred or disposed of;

[37] Requirements on the predisposal management of radioactive waste are established in Ref. [10] and for the disposal of radioactive waste are established in Ref. [11].

(f) Shall develop and implement a strategy for radioactive waste management and shall include appropriate evidence that protection and safety is optimized.

Discharges

3.132. Registrants and licensees, in cooperation with suppliers, in applying for an authorization for discharges, as appropriate:

(a) Shall determine the characteristics and activity of the material to be discharged, and the possible points and methods of discharge;
(b) Shall determine by an appropriate pre-operational study all significant exposure pathways by which discharged radionuclides could give rise to exposure of members of the public;
(c) Shall assess the doses to the representative person due to the planned discharges;
(d) Shall consider the radiological environmental impacts in an integrated manner with features of the system of protection and safety, as required by the regulatory body;
(e) Shall submit to the regulatory body the findings of (a) to (d) above as an input to the establishment by the regulatory body, in accordance with para. 3.123, of authorized limits on discharges and conditions for their implementation.

3.133. Registrants and licensees shall ensure that operational limits and conditions relating to public exposure are met in accordance with paras 3.123 and 3.124.

3.134. Registrants and licensees shall review and modify their discharge control measures, as appropriate and in agreement with the regulatory body, taking into account:

(a) Operating experience;
(b) Any changes in exposure pathways or in the characteristics of the representative person that could affect the assessment of doses due to the discharges.

Requirement 32: Monitoring and reporting

The regulatory body and relevant parties shall ensure that programmes for source monitoring and environmental monitoring are in place and that the results from the monitoring are recorded and are made available.

3.135. The regulatory body shall be responsible, as appropriate, for:

(a) Review and approval of monitoring programmes of registrants and licensees, which shall be sufficient for:
    (i) Verifying compliance with the requirements of these Standards in respect of public exposure in planned exposure situations;
    (ii) Assessing doses from public exposure;
(b) Review of periodic reports on public exposure (including results of monitoring programmes and dose assessments) submitted by registrants and licensees;
(c) Making provision for an independent monitoring programme;
(d) Assessment of the total public exposure from authorized sources and practices in the State on the basis of monitoring data provided by registrants and licensees and with the use of data from independent monitoring and assessments;
(e) Making provision for maintaining records of discharges, results of monitoring programmes and results of assessments of public exposure;
(f) Verification of compliance of an authorized practice with the requirements of these Standards for the control of public exposure.

3.136. The regulatory body shall publish or shall make available on request, as appropriate, results from source monitoring and environmental monitoring programmes and assessments of doses from public exposure.

3.137. Registrants and licensees shall, as appropriate:

(a) Establish and implement monitoring programmes to ensure that public exposure due to sources under their responsibility is adequately assessed and that the assessment is sufficient to verify and demonstrate compliance with the authorization. These programmes shall include monitoring of the following, as appropriate:
   - external exposure from such sources;
   - discharges;
   - radioactivity in the environment;
   - other parameters important for the assessment of public exposure.

(b) Maintain appropriate records of the results of the monitoring programmes and estimated doses to members of the public;

(c) Report or make available the results of the monitoring programme to the regulatory body at approved intervals, including, as applicable, the levels and composition of discharges, dose rates at the site boundary and in premises open to members of the public, results of environmental monitoring and retrospective assessments of doses to the representative person;

(d) Report promptly to the regulatory body any levels exceeding the operational limits and conditions relating to public exposure, including authorized limits on discharges, in accordance with reporting criteria established by the regulatory body;

(e) Report promptly to the regulatory body any significant increase in dose rate or concentrations of radionuclides in the environment that could be attributed to the authorized practice, in accordance with reporting criteria established by the regulatory body;

(f) Establish and maintain a capability to carry out monitoring in an emergency, in the event of unexpected increases in radiation levels or concentrations of radionuclides in the environment due to accidents or other unusual events attributed to the authorized source or facility;

(g) Verify the adequacy of the assumptions made for the assessment of public exposure and radiological environmental impacts;

(h) Publish or make available on request, as appropriate, results from source monitoring and environmental monitoring programmes and assessments of doses from public exposure.

Requirement 33: Consumer products

Providers of consumer products shall ensure that such products are not made available to the public unless their use by members of the public has been justified, and either their use has been exempted or their provision to the public has been authorized.

3.138. Providers of consumer products shall ensure that such products are not made available to the public unless the justification of their use by members of the public has been approved by the government or regulatory body, and either their use has been exempted on the basis of the criteria specified in Schedule I or their provision to the public has been authorized.

3.139. Upon receipt of a request for authorization to provide to the public consumer products, the regulatory body:

(a) Shall require the provider of the consumer product to provide documents to demonstrate compliance with the requirements in paras 3.138–3.143;

(b) Shall verify the assessments and the selection of parameters presented in the request for authorization;

(c) Shall determine whether the end use of the product can be exempted;

(d) Shall authorize the provision to the public of the consumer product, where appropriate, subject to specific conditions of authorization.

3.140. Providers of consumer products shall comply with the conditions of the authorization to provide such products to the public, shall ensure that such products comply with the requirements of these Standards, and shall plan for appropriate arrangements for the servicing, maintenance, recycling or disposal of such products. The design and manufacture of such products, with regard to features that could affect exposure during normal handling, transport and use, as well as in the event of mishandling, misuse, accident or disposal, shall be subject to optimization of protection and safety. In this regard, designers, manufacturers and other providers of consumer products shall take into account the following:

(a) The various radionuclides that could be used and their radiation types, energies, activities and half-lives;
(b) The chemical and physical forms of the radionuclides that could be used and their significance for protection and safety in normal conditions and abnormal conditions;
(c) The containment and shielding of radioactive substances in consumer products and access to these radioactive substances in normal conditions and abnormal conditions;
(d) The need for servicing or repair and ways in which this could be done;
(e) Relevant experience with similar consumer products.

3.141. Providers of consumer products shall ensure that:

(a) Where practicable, a legible label is firmly affixed to a visible surface of each such consumer product that:
    (i) States that the product contains radioactive substances and identifying the radionuclides and their activities;
    (ii) States that the provision of the product to the public has been authorized by the regulatory body;
    (iii) Provides information about required or recommended options for recycling or disposal;
(b) The information specified in (a) above is also printed legibly on the retail packaging of the consumer product.

3.142. Providers of consumer products shall provide clear and appropriate information and instructions with each such consumer product on:

(a) Correct installation, use and maintenance of the product;
(b) Servicing and repair;
(c) The radionuclides and their activities at a specified date;
(d) Dose rates in normal operation and during servicing and repair;
(e) Required or recommended options for recycling or disposal.

3.143. Providers of consumer products shall provide the product retailers with appropriate information on safety and instructions on transport and storage.

MEDICAL EXPOSURE

Scope

3.144. The requirements in respect of medical exposure in planned exposure situations (paras 3.144–3.184) apply to all medical exposures[38], including intended, unintended and accidental exposures.

3.145. Dose limits do not apply to medical exposures.

Requirement 34: Responsibilities of the government specific to medical exposure

The government shall ensure that relevant parties are authorized to assume their roles and responsibilities and that diagnostic reference levels, dose constraints, and criteria and guidelines for the release of patients are established.

3.146. The government, in accordance with paras 2.13–2.28, shall ensure with regard to medical exposures that, as a result of consultation between the health authority, relevant professional bodies and the regulatory body, the relevant parties identified in paras 2.40 and 2.41 are authorized to assume their roles and responsibilities, and shall ensure that they are notified of their duties in relation to protection and safety for individuals undergoing medical exposures.

3.147. The government shall ensure, as part of the responsibilities specified in para. 2.15, that as a result of consultation between the health authority, relevant professional bodies and the regulatory body, a set of diagnostic reference levels is established for medical exposures incurred in medical imaging, including image guided interventional procedures. In setting such diagnostic reference levels, account shall be taken of the need for adequate image quality, to enable the requirements of para. 3.168 to be fulfilled. Such diagnostic reference levels shall be based, as far as possible, on wide scale surveys or on published values that are appropriate for the local circumstances.

3.148. The government shall ensure that, as a result of consultation between the health authority, relevant professional bodies and the regulatory body, the following are established:

(a) Dose constraints, to enable the requirements of paras 3.172 and 3.173 respectively to be fulfilled for:
   (i) Exposures of carers and comforters[39];
   (ii) Exposures due to diagnostic investigations of volunteers participating in a programme of biomedical research;

(b) Criteria and guidelines for the release of patients who have undergone therapeutic procedures using unsealed sources or patients who still retain implanted sealed sources.

Requirement 35: Responsibilities of the regulatory body specific to medical exposure

The regulatory body shall require that health professionals with responsibilities for medical exposure are specialized in the appropriate area and that they meet the requirements for education, training and competence in the relevant specialty.

3.149. The regulatory body shall ensure that the authorization for medical exposures to be performed at a particular medical radiation facility allows personnel (radiological medical practitioners, medical physicists, medical radiation technologists and any other health professionals

---

[38] Requirements on human imaging for purposes other than medical diagnosis or treatment (and hence not within the scope of medical exposure) are stated in paras 3.61–3.67.

[39] The selection of constraints for carers and comforters is a complex process in which a number of factors have to be taken into account, such as the age of the individual and for a woman the possibility of her being pregnant.

with specific duties in relation to the radiation protection of patients) to take on the responsibilities specified in these Standards only if they:

(a) are specialized[40] in the appropriate area[41];
(b) meet the respective requirements for education, training and competence in radiation protection, in accordance with para. 2.32;
(c) are named in a list maintained up to date by the registrant or licensee.

Requirement 36: Responsibilities of registrants and licensees specific to medical exposure

Registrants and licensees shall ensure that no person incurs a medical exposure unless there has been an appropriate referral, responsibility has been assumed for ensuring protection and safety, and the person subject to exposure has been informed as appropriate of the expected benefits and risks.

3.150. Registrants and licensees shall ensure that no patient, whether symptomatic or asymptomatic, undergoes a medical exposure unless:

(a) The radiological procedure has been requested by a referring medical practitioner and information on the clinical context has been provided, or it is part of an approved health screening programme;
(b) The medical exposure has been justified through consultation between the radiological medical practitioner and the referring medical practitioner, as appropriate, or it is part of an approved health screening programme;
(c) A radiological medical practitioner has assumed responsibility for protection and safety in the planning and delivery of the medical exposure as specified in para. 3.153(a);
(d) The patient or the patient's legal authorized representative has been informed, as appropriate, of the expected diagnostic or therapeutic benefits of the radiological procedure as well as the radiation risks.

3.151. Registrants and licensees shall ensure that no individual incurs a medical exposure as part of a programme of biomedical research unless the exposure has been approved by an ethics committee (or other institutional body that has been assigned similar functions by the relevant authority) as required in para. 3.160 and a radiological medical practitioner has assumed responsibility as specified in para. 3.153(a). Registrants and licensees shall ensure that the requirements specified in para. 3.173 are met for the optimization of protection and safety for persons subject to exposure as part of a programme of biomedical research.

3.152. Registrants and licensees shall ensure that no individual incurs a medical exposure as a carer or comforter unless he or she has received, and has indicated an understanding of, relevant information on radiation protection and information on the radiation risks prior to providing care and comfort to an individual undergoing a radiological procedure. Registrants and licensees shall ensure that the requirements specified in para. 3.172 are met for the optimization of protection and safety for any procedure in which an individual acts as a carer or comforter.

3.153. Registrants and licensees shall ensure that:

---

[40] 'Specialized' mean as acknowledged by the relevant professional body, health authority or appropriate organization.

[41] 'The appropriate area' means, in the first instance, diagnostic radiology, image guided interventional procedures, or radiotherapy or nuclear medicine (diagnostic, therapeutic or both). Often, the area of specialization is likely to be narrower, however, in particular with regard to the radiological medical practitioner. Examples are dental, chiropractic, or podiatric specialists in the case of diagnostic radiology, and cardiologists, urologists or neurologists in the case of image guided interventional procedures.

(a) The radiological medical practitioner performing or overseeing the radiological procedure has assumed responsibility for ensuring overall protection and safety for patients during the planning and delivery of the medical exposure, including the justification of the procedure as required in paras 3.154–3.160 and the optimization of protection and safety, in cooperation with the medical physicist and the medical radiation technologist as required in paras 3.161–3.176;
(b) Radiological medical practitioners, medical physicists, medical radiation technologists and other health professionals with specific duties in relation to protection and safety for patients in a given radiological procedure have the appropriate specialization;
(c) Sufficient medical personnel and paramedical personnel are available as specified by the health authority;
(d) For therapeutic uses of radiation, the requirements of these Standards for calibration, dosimetry and quality assurance, including the acceptance and commissioning of medical radiological equipment, as specified in paras 3.166, 3.167(c), 3.169 and 3.170, are fulfilled by or under the supervision of a medical physicist;
(e) For diagnostic radiological procedures and image guided interventional procedures, the requirements of these Standards for medical imaging, calibration, dosimetry and quality assurance, including the acceptance and commissioning of medical radiological equipment, as specified in paras 3.166, 3.167(a), 3.167(b), 3.168, 3.169 and 3.170, are fulfilled by or under the oversight of or with the documented advice of a medical physicist, whose degree of involvement is determined by the complexity of the radiological procedures and the associated radiation risks;
(f) Any delegation of responsibilities by a principal party is documented.

Requirement 37: Justification of medical exposures

Relevant parties shall ensure that medical exposures are justified.

3.154. Medical exposures shall be justified by weighing the expected diagnostic or therapeutic benefits[42] that they yield against the radiation detriment that they might cause, with account taken of the benefits and the risks of available alternative techniques that do not involve medical exposure.

3.155. Generic justification of a radiological procedure shall be carried out by the health authority in conjunction with appropriate professional bodies, and shall be reviewed from time to time, with account taken of advances in knowledge and technological developments.

3.156. The justification of medical exposure for an individual patient shall be carried out through consultation between the radiological medical practitioner and the referring medical practitioner, as appropriate, with account taken, in particular for patients who are pregnant or breast-feeding or paediatric, of:

(a) The appropriateness of the request;
(b) The urgency of the procedure;
(c) The characteristics of the medical exposure;
(d) The characteristics of the individual patient;
(e) Relevant information from the patient's previous radiological procedures.

3.157. Relevant national or international referral guidelines shall be taken into account for the justification of the medical exposure of an individual patient in a radiological procedure.

---

[42] The benefit may not necessarily be to the person exposed. Clearly for patients this is the case, but for exposures in biomedical research the benefit is expected to be for biomedical sciences and for future health care. Similarly, the benefit for carers and comforters might be, for example, the successful performance of a diagnostic procedure on a child.

3.158. Justification for radiological procedures to be performed as part of a health screening programme for asymptomatic populations shall be carried out by the health authority in conjunction with appropriate professional bodies.

3.159. Any radiological procedure on an asymptomatic individual that is intended to be performed for the early detection of disease, but not as part of an approved health screening programme, shall require specific justification for that individual by the radiological medical practitioner and the referring medical practitioner, in accordance with the guidelines of relevant professional bodies or the health authority. As part of this process, the individual shall be informed in advance of the expected benefits, risks and limitations of the procedure.

3.160. The medical exposure of volunteers as part of a programme of biomedical research is deemed to be not justified unless:

(a) It is in accordance with the provisions of the Helsinki Declaration [20] and takes into account the guidelines published by the Council for International Organizations of Medical Sciences [21], together with the recommendations of the ICRP [22];

(b) It is subject to approval by an ethics committee (or other institutional body that has been assigned similar functions by the relevant authority), subject to any dose constraints that may be specified (as required in paras 3.148(a)(ii) and 3.173), and subject to applicable national regulations and local regulations.

Requirement 38: Optimization of protection and safety

Registrants and licensees and radiological medical practitioners shall ensure that protection and safety is optimized for each medical exposure.

Design considerations

3.161. In addition to ensuring that the responsibilities stated in para. 3.49 are discharged, as applicable, registrants and licensees, in cooperation with suppliers, shall ensure that medical radiological equipment, and software that could influence the delivery of medical exposure is used only if it conforms to the applicable standards of the International Electrotechnical Commission and the International Organization for Standardization or to national standards adopted by the regulatory body.

Operational considerations

3.162. For diagnostic radiological procedures and image guided interventional procedures, the radiological medical practitioner, in cooperation with the medical radiation technologist and the medical physicist, and if appropriate with the radiopharmacist or radiochemist , shall ensure that the following are used:

(a) Appropriate medical radiological equipment and software and also, for nuclear medicine, appropriate radiopharmaceuticals;

(b) Appropriate techniques and parameters to deliver a medical exposure of the patient that is the minimum necessary to fulfil the clinical purpose of the procedure, with account taken of relevant norms of acceptable image quality established by relevant professional bodies and relevant diagnostic reference levels established in accordance with paras 3.147 and 3.168.

3.163. For therapeutic radiological procedures, the radiological medical practitioner, in cooperation with the medical physicist and the medical radiation technologist, shall ensure that for each patient the exposure of volumes other than the planning target volume is kept as low as reasonably achievable consistent with delivery of the prescribed dose to the planning target volume within the required tolerances.

3.164. For therapeutic radiological procedures in which radiopharmaceuticals are administered, the radiological medical practitioner, in cooperation with the medical physicist and the medical radiation technologist, and if appropriate with the radiopharmacist or radiochemist, shall ensure that for each patient the appropriate radiopharmaceutical with the appropriate activity is selected and administered so that the radioactivity is primarily localized in the organ(s) of interest, while the radioactivity in the rest of the body is kept as low as reasonably achievable.

3.165. Registrants and licensees shall ensure that the particular aspects of medical exposures are considered in the optimization process for:

(a) Paediatric patients subject to medical exposure;
(b) Individuals subject to medical exposure as part of a health screening programme;
(c) Volunteers subject to medical exposure as part of a programme of biomedical research;
(d) Relatively high doses[43] to the patient;
(e) Exposure of the embryo or fetus, in particular for radiological procedures in which the abdomen or pelvis of the pregnant woman is exposed to the useful radiation beam or could otherwise receive a significant dose;
(f) Exposure of a breast-fed infant as a result of a female patient undergoing a radiological procedure with radiopharmaceuticals.

Calibration

3.166. In accordance with para. 3.153(d) and (e), the medical physicist shall ensure that:

(a) All sources giving rise to medical exposure are calibrated in terms of appropriate quantities using internationally accepted or nationally accepted protocols;
(b) Calibrations are carried out at the time of commissioning a unit prior to clinical use, after any maintenance procedure that could affect the dosimetry and at intervals approved by the regulatory body;
(c) Calibrations of radiotherapy units are subject to independent verification[44] prior to clinical use;
(d) Calibration of all dosimeters used for dosimetry of patients and for the calibration of sources is traceable to a standards dosimetry laboratory.

Dosimetry of patients

3.167. Registrants and licensees shall ensure that dosimetry of patients is performed and documented by or under the supervision of a medical physicist, using calibrated dosimeters and following internationally accepted or nationally accepted protocols, including dosimetry to determine the following:

(a) For diagnostic medical exposures, typical doses to patients for common radiological procedures;
(b) For image guided interventional procedures, typical doses to patients;

---

[43] The term 'relatively high dose' is intended to apply in a given context. Clearly doses from therapeutic exposures are included in 'relatively high doses', as are image guided interventional procedures. In diagnostic medical imaging, 'relatively high doses' would include doses from exposures in computerized tomography and the procedures with higher doses in nuclear medicine.

[44] 'Independent verification' ideally means verification by a different, independent medical physicist using different dosimetry equipment. However, other options, such as verification by a second medical physicist or only verification using a second set of equipment, or even using a form of verification by postal thermoluminescence dosimetry could be acceptable. In checking for compliance, the regulatory body needs to be aware of the limitations on local resources.

(c) For therapeutic medical exposures, absorbed doses to the tissues or organs for individual patients, as determined to be relevant by the radiological medical practitioner.

Diagnostic reference levels

3.168. Registrants and licensees shall ensure that:

(a) Local assessments, on the basis of the measurements required in para. 3.167, are made at approved intervals for those radiological procedures for which diagnostic reference levels have been established (para. 3.147);
(b) A review is conducted to determine whether the optimization of protection and safety for patients is adequate, or whether corrective action is required if, for a given radiological procedure:
    (i) typical doses or activities exceed the relevant diagnostic reference level; or
    (ii) typical doses or activities fall substantially below the relevant diagnostic reference level and the exposures do not provide useful diagnostic information or do not yield the expected medical benefit to the patient.

Quality assurance for medical exposures

3.169. Registrants and licensees, in applying the requirements of these Standards in respect of management systems, shall establish a comprehensive programme of quality assurance for medical exposures with the active participation of medical physicists, radiological medical practitioners, medical radiation technologists and, for complex nuclear medicine facilities, radiopharmacists and radiochemists, and in conjunction with other health professionals as appropriate. Principles established by the World Health Organization, the Pan American Health Organization and relevant professional bodies shall be taken into account.

3.170. Registrants and licensees shall ensure that programmes of quality assurance for medical exposure include, as appropriate to the medical radiation facility:

(a) Measurements of the physical parameters of medical radiological equipment made by, or under the supervision of, a medical physicist:
    (i) At the time of acceptance and commissioning of the equipment prior to its clinical use on patients;
    (ii) Periodically thereafter;
    (iii) After any major maintenance procedure that could affect protection and safety of patients;
    (iv) After any installation of new software or modification of existing software that could affect protection and safety of patients;
(b) Implementation of corrective actions if measured values of the physical parameters mentioned in (a) are outside established tolerance limits;
(c) Verification of the appropriate physical and clinical factors used in radiological procedures;
(d) Maintaining records of relevant procedures and results;
(e) Periodic checks of the calibration and conditions of operation of dosimetry equipment and monitoring equipment.

3.171. Registrants and licensees shall ensure that regular and independent audits are made of the programme of quality assurance for medical exposures, and that their frequency is in accordance with the complexity of the radiological procedures being performed and the associated risks.

Dose constraints

3.172. Registrants and licensees shall ensure that relevant dose constraints (para. 3.148(a)(i)) are used in the optimization of protection and safety in any procedure in which an individual acts as a carer or comforter.

3.173. Registrants and licensees shall ensure that dose constraints specified or approved by the ethics committee, or by another institutional body that has been assigned similar functions by the relevant authority, on a case by case basis as part of a proposal for biomedical research (para. 3.160) are used in the optimization of protection and safety for persons subject to exposure as part of a programme of biomedical research.

Requirement 39: Pregnant women and breast-feeding women

Registrants and licensees shall ensure that there are arrangements in place for appropriate radiation protection in cases where a woman is or might be pregnant or is breast-feeding.

3.174. Registrants and licensees shall ensure that signs in appropriate languages are placed in public places, waiting rooms for patients, cubicles and other appropriate places, and that other means of communication are also used as appropriate[45], to request female patients who are to undergo a radiological procedure to notify the radiological medical practitioner, medical radiation technologist or other personnel in the event that:

(a) She is or she might be pregnant;
(b) She is breast-feeding and the scheduled radiological procedure includes the administration of a radiopharmaceutical.

3.175. Registrants and licensees shall ensure that there are procedures in place for ascertaining the pregnancy status of a female patient of reproductive capacity before the performance of any radiological procedure that could result in a significant dose to the embryo or fetus, so that this information can be considered in the justification for the radiological procedure (paras. 3.154 and 3.156) and in the optimization of protection and safety (para. 3.165).

3.176. Registrants and licensees shall ensure that there are arrangements in place for establishing that a female patient is not breast-feeding before the performance of any radiological procedure involving the administration of a radiopharmaceutical that could result in a significant dose to an infant being breast-fed, so that this information can be considered in the justification for the radiological procedure (paras 3.154 and 3.156) and in the optimization of protection and safety (para. 3.165).

Requirement 40: Release of patients after radionuclide therapy

Registrants and licensees shall ensure that there are arrangements in place to ensure appropriate radiation protection for members of the public and for family members before a patient is released following radionuclide therapy.

3.177. The radiological medical practitioner shall ensure that no patient who has undergone a therapeutic procedure with sealed sources or unsealed sources is discharged from a medical radiation facility until it has been established by either a medical physicist or the facility's radiation protection officer that:

---

[45] This includes explicitly asking female patients whether they are or might be pregnant or are breast-feeding.

(a) The activity of radionuclides in the patient is such that doses that could be received by members of the public and family members would be in compliance with the requirements set by the relevant authorities (para. 3.148(b)); and
(b) The patient or legal guardian of the patient is provided with:
    (i) Written instructions for keeping doses to persons in contact with or in the vicinity of the patient as low as reasonably achievable and for avoiding the spread of contamination;
    (ii) Information on the radiation risks.

Requirement 41: Unintended and accidental medical exposures

Registrants and licensees shall ensure that all practicable measures are taken to minimize the likelihood of unintended or accidental medical exposures. Registrants and licensees shall promptly investigate any such exposure and, if appropriate, shall implement corrective actions.

3.178. Registrants and licensees, in accordance with the relevant requirements of paras 2.51, 3.41–3.44 and 3.50, shall ensure that all practicable measures are taken to minimize the likelihood of unintended or accidental medical exposures arising from flaws in design and operational failures of medical radiological equipment, from failures of and errors in software, or as a result of human error.

Investigation of unintended and accidental medical exposures

3.179. Registrants and licensees shall promptly investigate any of the following unintended or accidental medical exposures:

(a) Any medical treatment delivered to the wrong individual or to the wrong tissue of the patient, or using the wrong radiopharmaceutical, or with an activity, a dose or dose fractionation differing substantially from (over or under) the values prescribed by the radiological medical practitioner, or that could lead to unduly severe secondary effects;
(b) Any diagnostic radiological procedure or image guided interventional procedure in which the wrong individual or the wrong tissue of the patient is subject to exposure;
(c) Any exposure for diagnostic purposes that is substantially greater than was intended;
(d) Any exposure arising from an image guided interventional procedure that is substantially greater than was intended;
(e) Any inadvertent exposure of the embryo or fetus in the course of performing a radiological procedure;
(f) Any failure of medical radiological equipment, software failure or system failure, or accident, error, mishap or other unusual occurrence with the potential for subjecting the patient to a medical exposure that is substantially different from what was intended.

3.180. Registrants and licensees shall, with regard to any unintended or accidental medical exposures investigated as required in para. 3.179:

(a) Calculate or estimate the doses received and the dose distribution within the patient;
(b) Indicate the corrective actions required to prevent the recurrence of such an unintended or accidental medical exposure;
(c) Implement all the corrective actions that are under their own responsibility;
(d) Produce and keep, as soon as possible after the investigation or as otherwise required by the regulatory body, a written record that states the cause of the unintended or accidental medical exposure and includes the information specified in (a) to (c) above, as relevant, and any other information as required by the regulatory body; and for significant unintended or accidental medical exposures or as otherwise required by the regulatory body, submit this written record, as soon as possible, to the regulatory body, and to the relevant health authority if appropriate;
(e) Ensure that the appropriate radiological medical practitioner informs the referring medical practitioner and the patient or the patient's legal authorized representative of the unintended or accidental medical exposure.

Requirement 42: Reviews and records

Registrants and licensees shall ensure that radiological reviews are performed periodically at medical radiation facilities and that records are maintained.

Radiological reviews

3.181. Registrants and licensees shall ensure that radiological reviews are performed periodically by the radiological medical practitioners at the medical radiation facility, in cooperation with the medical radiation technologists and the medical physicists. The radiological review shall include an investigation and critical review of the current practical application of the radiation protection principles of justification and optimization for the radiological procedures that are performed in the medical radiation facility.

Records

3.182. Registrants and licensees shall maintain for a period as specified by the regulatory body and shall make available, as required, the following personnel records:

(a) Records of any delegation of responsibilities by principal parties (as required in para. 3.153(f));
(b) Records of training of personnel in radiation protection (as required in para. 3.149(b)).

3.183. Registrants and licensees shall maintain for a period as specified by the regulatory body and shall make available, as required, the following records of calibration, dosimetry and quality assurance:

(a) Records of the results of the calibrations and periodic checks of the relevant physical and clinical parameters selected during treatment of patients;
(b) Records of dosimetry of patients, as required in para. 3.167;
(c) Records of local assessments and reviews made with regard to diagnostic reference levels, as required in para. 3.168;
(d) Records associated with the quality assurance programme, as required in para. 3.170(d).

3.184. Registrants and licensees shall maintain for a period as specified by the regulatory body and shall make available, as required, the following records for medical exposure:

(a) For diagnostic radiology, information necessary for retrospective assessment of doses, including the number of exposures and the duration of fluoroscopic radiological procedures;
(b) For image guided interventional procedures, information necessary for retrospective assessment of doses, including the duration of the fluoroscopic component and the number of images acquired;
(c) For nuclear medicine, the types of radiopharmaceutical administered and their activity;
(d) For radiation therapy, a description of the planning target volume, the dose to the centre of the planning target volume, and the maximum and minimum doses delivered to the planning target volume, or equivalent alternative information on doses to the planning target volume, the doses to relevant organs as selected by the radiological medical practitioner, the dose fractionation, and the overall treatment time;
(e) Exposure records for volunteers subject to medical exposure as part of a programme of biomedical research;
(f) Reports on investigations of unintended and accidental medical exposures (as required in para. 3.180(d)).

4. EMERGENCY EXPOSURE SITUATIONS

SCOPE

4.1. The requirements for emergency exposure situations established in Section 4 apply to activities undertaken in preparedness for and in response to a nuclear or radiation emergency.

GENERIC REQUIREMENTS

Requirement 43: Emergency management system

The government shall ensure that an integrated and coordinated emergency management system is established and maintained.

4.2. The government shall ensure that an emergency management system is established and maintained on the territories and within the jurisdiction of the State for the purposes of emergency response to protect human life, health and the environment in the event of a nuclear or radiation emergency.

4.3. The emergency management system shall be designed to be commensurate with the results of a hazard assessment [15] and to enable an effective emergency response to reasonably foreseeable events (including very low probability events) in connection with facilities or activities.

4.4. The emergency management system shall be integrated, to the extent practicable, into an all-hazards emergency management system.

4.5. The emergency management system shall provide for essential elements at the scene, and at the local, national and international level, as appropriate, including the following [15]:

(a) Hazard assessment;
(b) Development and exercising of emergency plans and emergency procedures;
(c) Clear allocation of responsibilities of persons and organizations having roles in the arrangements for emergency preparedness and response;
(d) Arrangements for efficient and effective cooperation and coordination among organizations;
(e) Reliable communication, including public information;
(f) Optimized protection strategies for the implementation and the termination of measures for the protection of members of the public who could be subject to exposure in an emergency, including relevant considerations for protection of the environment;
(g) Arrangements for the protection of emergency workers;
(h) Education and training, including training in radiation protection, of all persons involved in emergency response and exercising of emergency plans and emergency procedures;
(i) Preparations for the transition from an emergency exposure situation to an existing exposure situation;
(j) Arrangements for the medical response and the public health response in an emergency;
(k) Provision for individual monitoring and environmental monitoring and for dose assessment;
(l) Involvement of relevant parties and interested parties.

4.6. The government shall ensure the coordination of its emergency arrangements and capabilities with international emergency arrangements.

PUBLIC EXPOSURE

Requirement 44: Preparedness and response to an emergency

The government shall ensure that protection strategies are developed, justified and optimized at the planning stage, and that emergency response is undertaken through their timely implementation.

4.7. The government shall ensure that protection strategies are developed, justified, and optimized at the planning stage, by using scenarios based on the hazard assessment, for avoiding deterministic effects and reducing the likelihood of stochastic effects due to public exposure.

4.8. Development of a protection strategy shall include, but shall not be limited to, the following three successive steps:

(1) A reference level expressed in terms of residual dose shall be set, typically an effective dose in the range 20–100 mSv, that includes dose contributions via all exposure pathways. The protection strategy shall include planning for residual doses to be as low as reasonably achievable below the reference level, and the strategy shall be optimized.

(2) On the basis of the outcome of the optimization of the protection strategy, using the reference level, generic criteria for particular protective actions and other actions, expressed in terms of projected dose or dose that has been received, shall be developed. If the numerical values of the generic criteria[46] are exceeded, those protective actions and other actions, either individually or in combination, shall be implemented.

(3) Once the protection strategy has been optimized and a set of generic criteria has been developed, pre-established default triggers for initiating the different parts of an emergency plan, primarily for the initial phase, shall be derived from the generic criteria. Default triggers, such as on-scene conditions, operational intervention levels and emergency action levels, shall be expressed in terms of parameters or observable conditions. Arrangements shall be established in advance to revise these triggers, as appropriate, in an emergency exposure situation, with account taken of the prevailing conditions as these evolve.

4.9. Each protective action shall be justified in the context of the protection strategy.

4.10. The government shall ensure that in making arrangements for emergency preparedness and response it is taken into consideration that emergencies are dynamic, that decisions taken early in the response may have an impact on subsequent actions, and that different geographical areas may have different prevailing conditions and different requirements for the response.

4.11. The government shall ensure that the response in an emergency exposure situation is undertaken through the timely implementation of arrangements for emergency response, including but not limited to:

(a) Promptly implementing protective actions to avoid severe deterministic effects on the basis of observed conditions and, if possible, before any exposure occurs. Dose levels required to be used as generic criteria for preventing severe deterministic effects are given in Schedule IV, Table IV-1;

(b) Assessing the effectiveness of the actions implemented and modifying them as appropriate;

(c) Comparing residual doses with the applicable reference level, giving priority to those groups for whom residual doses exceed the reference level;

---

[46] Table A-1 in the Annex provides a set of generic criteria for use in the protection strategy that are compatible with reference levels within a range of 20–100 mSv, and further details for specific actions in different time frames.

(d) Implementing further protection strategies as necessary, on the basis of prevailing conditions and available information.

EXPOSURE OF EMERGENCY WORKERS

Requirement 45: Arrangements for controlling the exposure of emergency workers

The government shall establish a programme for managing, controlling and recording the doses received in an emergency by emergency workers.

4.12. The government shall establish a programme for managing, controlling and recording the doses received in an emergency by emergency workers, which shall be implemented by response organizations and employers.

4.13. The response organization and employers responsible for ensuring compliance with the requirements in paras 4.14–4.19 shall be specified in the emergency plan.

4.14. In an emergency exposure situation, the relevant requirements for occupational exposure in planned exposure situations (paras 3.68–3.116) shall be applied for emergency workers, in accordance with a graded approach, except as required in para. 4.15.

4.15. Response organizations and employers shall ensure that no emergency worker is subject to an exposure in an emergency in excess of 50 mSv other than:

(a) For the purposes of saving life or preventing serious injury;
(b) When undertaking actions to prevent severe deterministic effects and actions to prevent the development of catastrophic conditions that could significantly affect people and the environment; or
(c) When undertaking actions to avert a large collective dose.

4.16. In the exceptional circumstances of para. 4.15, response organizations and employers shall make all reasonable efforts to keep doses to emergency workers below the values set out in Schedule IV, Table IV-2. In addition, emergency workers undertaking actions due to which their doses could approach or exceed the values set out in Schedule IV, Table IV-2 shall do so only when the expected benefits to others would clearly outweigh the risks to the emergency workers.

4.17. Response organizations and employers shall ensure that emergency workers who undertake actions in which the doses received might exceed 50 mSv do so voluntarily[47]; that they have been clearly and comprehensively informed in advance of the associated health risks, as well as of available protective measures; and that they are, to the extent possible, trained in the actions that they may be required to take.

4.18. Response organizations and employers shall take all reasonable steps to assess and record the doses received in an emergency by emergency workers. Information of the doses received and information concerning the associated health risks shall be communicated to the workers involved.

4.19. Workers who receive doses in an emergency exposure situation shall not normally be precluded from incurring further occupational exposure. However, qualified medical advice shall be obtained before any further occupational exposure if a worker has received a dose exceeding 200 mSv or at the request of the worker.

---

[47] The voluntary basis for response actions by emergency workers is usually covered in the emergency response arrangements.

TRANSITION FROM AN EMERGENCY EXPOSURE SITUATION TO AN EXISTING EXPOSURE SITUATION

Requirement 46: Arrangements for the transition from an emergency exposure situation to an existing exposure situation

The government shall ensure that arrangements are in place and are implemented as appropriate for the transition from an emergency exposure situation to an existing exposure situation.

4.20. The government shall ensure that, as part of its overall emergency preparedness, arrangements are in place for the transition from an emergency exposure situation to an existing exposure situation. The arrangements shall take into account that different geographic areas may undergo the transition at different times. The responsible authority shall take the decision to make the transition to an existing exposure situation. The transition shall be made in a coordinated and orderly manner, by making any necessary transfer of responsibilities between organizations, with the involvement of relevant authorities and interested parties.

4.21. Workers undertaking work such as repairs to plant and buildings or activities for radioactive waste management or remedial work for the decontamination of the site and surrounding areas, shall be subject to the relevant requirements for occupational exposure in planned exposure situations stated in Section 3.

5. EXISTING EXPOSURE SITUATIONS

SCOPE

5.1. The requirements for existing exposure situations in Section 5 apply to:

(a) Exposure due to contamination of areas by residual radioactive material arising from:
   (i) Past activities that were never subject to regulatory control or that were subject to regulatory control but not in accordance with the requirements of these Standards;
   (ii) A nuclear or radiation emergency, after an emergency exposure situation has been declared ended (as required in para. 4.20);

(b) Exposure due to commodities, including food, feed, drinking water and construction materials, that incorporate radionuclides arising from residual radioactive material as stated in para. 5.1(a);

(c) Exposure due to natural sources, including:
   (i) ^{222}Rn, ^{220}Rn and their progeny, in workplaces other than those for which exposure due to other radionuclides in the uranium or thorium decay chains is controlled as a planned exposure situation, in dwellings and in other buildings with high occupancy factors for members of the public;
   (ii) Radionuclides of natural origin, regardless of activity concentration, in commodities, including food, feed, drinking water, agricultural fertilizer and soil amendments, and construction material, and existing residues in the environment;
   (iii) Materials, other than those stated in para. 5.1(c)(ii), in which the activity concentration of no radionuclide in either the uranium or thorium decay chains exceeds 1 Bq/g or the activity concentration of ^{40}K does not exceed 10 Bq/g;
   (iv) Exposure of aircrew and space crew to cosmic radiation.

GENERIC REQUIREMENTS

Requirement 47: Responsibilities of the government specific to existing exposure situations

The government shall ensure that existing exposure situations that have been identified are evaluated to determine which occupational exposures and public exposures are of concern from the point of view of radiation protection.

5.2. The government shall ensure that, when an existing exposure situation is identified, responsibilities for protection and safety are assigned and appropriate reference levels are established.

5.3. The government shall include in the legal and regulatory framework for protection and safety (see Section 2) provision for the management of existing exposure situations. The government, in the legal and regulatory framework, as appropriate:

(a) Shall specify the exposure situation that are included in the scope of existing exposure situations;[48]

---

[48] In the case of exposure due to radon, the types of situation that are included in the scope of existing exposure situations will include exposure in workplaces for which the exposure due to radon is not required by or directly related to the work and for which annual average activity concentrations due to ^{222}Rn might be expected to exceed the reference level established in accordance with para. 5.27.

(b) Shall specify the general principles underlying the protection strategies developed to reduce exposure when remedial actions and protective actions have been determined to be justified;[49]

(c) Shall assign responsibilities for the establishment and implementation of protection strategies to the regulatory body and to other relevant authorities[50] and, as appropriate, to registrants, licensees and other parties involved in the implementation of remedial and protective actions;

(d) Shall provide for the involvement of interested parties in decisions regarding the development and implementation of protection strategies, as appropriate.

5.4. The regulatory body or other relevant authority assigned to establish a protection strategy for an existing exposure situation shall ensure that it defines:

(a) The objectives to be achieved by means of the protection strategy;

(b) Appropriate reference levels.

5.5. The regulatory body or other relevant authority shall implement the protection strategy, including:

(a) Arranging for evaluation of the available remedial actions and protective actions for achieving the objectives, and for evaluation of the efficiency of the actions planned and implemented;

(b) Ensuring that information is available to individuals subject to exposure on potential health risks and on the means available for reducing their exposures and the associated risks.

PUBLIC EXPOSURE

Scope

5.6. The requirements in respect of public exposure in existing exposure situations (paras 5.7–5.23) apply to any public exposure arising from the situations specified in para. 5.1.

Requirement 48: Justification for protective actions and optimization of protection and safety

The government and the regulatory body or other relevant authority shall ensure that remedial actions and protective actions are justified and that the protection and safety is optimized.

5.7. The government and the regulatory body or other relevant authority shall ensure that the protection strategy for the management of existing exposure situations, established in accordance with paras 5.2 and 5.4, is commensurate with the radiation risks associated with the existing exposure situation; and that remedial actions or protective actions are expected to yield sufficient benefits to outweigh the detriments associated with taking them, including detriments in the form of radiation risks.[51]

5.8. The regulatory body or other relevant authority and other parties responsible for remedial actions or protective actions shall ensure that the form, scale and duration of such actions are

---

[49] Such actions include remedial actions such as the removal or reduction of the source giving rise to the exposure, as well as other longer term protective actions such as restriction of the use of construction materials, restriction of the consumption of foodstuffs and restriction of land use or of access to land or buildings.

[50] In existing exposure situations that do not fall under the jurisdiction of the regulatory body, another relevant authority, such as a health authority, may have the authority for implementing measures for protection and safety.

[51] The implementation of remedial actions (remediation) does not imply the elimination of all radioactivity or all traces of radioactive material. The optimization process may lead to extensive remediation but not necessarily to the restoration of previous conditions.

optimized. While this optimization process is intended to provide optimized protection for all individuals subject to exposure, priority shall be given to those groups for whom residual dose exceeds the reference level. All reasonable steps shall be taken to prevent doses remaining above the reference levels. Reference levels shall typically be expressed as an annual effective dose to the representative person in the range 1–20 mSv or other equivalent quantity, the actual value depending on the feasibility of controlling the situation and experience in managing similar situations in the past.

5.9. The regulatory body or other relevant authority shall periodically review the reference levels to ensure that they remain appropriate in the light of the prevailing circumstances.

Requirement 49: Responsibilities for remediation of areas with residual radioactive material

The government shall ensure that provision is made for identifying those persons or organizations responsible for areas with residual radioactive material, for establishing and implementing remediation programmes and post-remediation control measures, if appropriate, and for putting in place an appropriate strategy for radioactive waste management.

5.10. For the remediation of areas with residual radioactive material from past activities or from a nuclear or radiation emergency (para. 5.1(a)), the government shall ensure that provision is made in the framework for protection and safety for:

(a) The identification of those persons or organizations responsible for the contamination of areas and those responsible for financing the remediation programme, and the determination of appropriate arrangements for alternative sources of funding if such persons or organizations are no longer present or are unable to meet their liabilities;
(b) The designation of persons or organizations responsible for planning, implementing and verifying the results of remedial actions;
(c) The establishment of any restrictions on the use of or access to the areas concerned before, during and, if necessary, after remediation;
(d) An appropriate system for maintaining, retrieval and amendment of records that cover the nature and the extent of contamination; the decisions made before, during and after remediation; and information on verification of the results of remedial actions, including the results of all monitoring and surveillance programmes after completion of the remedial actions.

5.11. The government shall ensure that a strategy for radioactive waste management is put in place to deal with any waste arising from the remedial actions and that provision for such a strategy is made in the framework for protection and safety.

5.12. The persons or organizations responsible for the planning, implementation and verification of remedial actions shall, as appropriate, ensure that:

(a) A remedial action plan, supported by a safety assessment, is prepared and is submitted to the regulatory body or other relevant authority for approval;
(b) The remedial action plan is aimed at the timely and progressive reduction of the radiation risks and eventually, if possible, the removal of restrictions on use of or access to the area;
(c) Any additional dose received by members of the public as a result of the remedial actions is justified on the basis of the resulting net benefit, including consideration of the consequent reduction of the annual dose;
(d) In the choice of the optimized remediation option:
    (i) The radiological impacts on people and the environment are considered together with non-radiological impacts on people and the environment, and technical, societal and economic factors;

(ii) The costs of the transport and management of radioactive waste, the radiation exposure of and health risks to the workers managing the waste, and any subsequent public exposure associated with its disposal are all taken into account;

(e) A mechanism for public information is in place and the interested parties affected by the existing exposure situation are involved in the planning, implementation and verification of the remedial actions, including any monitoring and surveillance following remediation;

(f) A monitoring programme is established and implemented;

(g) A system for maintaining adequate records relating to the existing exposure situation and actions taken for protection and safety is in place;

(h) Procedures are in place for reporting to the regulatory body on any abnormal conditions relevant to protection and safety.

5.13. The regulatory body or other relevant authority, in accordance with para. 2.29, shall take responsibility in particular for:

(a) Review of the safety assessment submitted by the responsible person or organization, approval of the remedial action plan and of any subsequent changes to the remedial action plan, and granting of any necessary authorization;

(b) Establishment of criteria and methods for assessing safety;

(c) Review of work procedures, monitoring programmes and records;

(d) Review and approval of significant changes to procedures or equipment that may have radiological environmental impacts or that may alter the exposure conditions for workers taking remedial actions or members of the public;

(e) Where necessary, establishment of regulatory requirements for control measures following remediation.

5.14. The person or organization responsible for carrying out the remedial actions:

(a) Shall ensure that the work, including management of the radioactive waste arising, is conducted in accordance with the remedial action plan;

(b) Shall take responsibility for all aspects of protection and safety, including the performance of a safety assessment;

(c) Shall monitor and perform a radiological survey of the area regularly during the remediation work so as to verify levels of contamination, to verify compliance with the requirements for waste management, and to enable any unexpected levels of radiation to be detected and the remedial action plan to be modified accordingly, subject to approval by the regulatory body or other relevant authority;

(d) Shall perform a radiological survey after completion of remedial actions to demonstrate that the end point conditions, as established in the remedial action plan, have been met;

(e) Shall prepare and retain a final remediation report and shall submit a copy to the regulatory body or other relevant authority.

5.15. After the remedial actions have been completed, the regulatory body or other relevant authority:

(a) Shall review, amend as necessary and formalize the type, extent and duration of any post-remediation control measures already identified in the remedial action plan, with due consideration of the residual radiation risks;

(b) Shall identify the person or organization responsible for any post-remediation control measures;

(c) Shall where necessary impose specific restrictions for the remediated area to control:

   (i) Access by unauthorized persons;

   (ii) Removal of radioactive material or use of such material, including its use in commodities;

(iii) Future use of the area, including the use of water resources and use for the production of food or feed, and the consumption of food from the area;

(d) Shall periodically review conditions in the remediated area and, if appropriate, shall amend or remove any restrictions.

5.16. The person or organization responsible for post-remediation control measures shall establish and maintain for as long as required by the regulatory body or other relevant authority an appropriate programme, including any necessary provisions for monitoring and surveillance, to verify the long term effectiveness of the completed remedial actions for areas in which controls are required after remediation has been completed.

5.17. For those areas with long lasting residual radioactive material in which the government has decided to allow habitation and the resumption of social and economic activities, the government, in consultation with interested parties, shall ensure that arrangements are in place, as necessary, for the ongoing control of exposure with the aim of establishing conditions for sustainable living, including:

(a) Establishment of reference levels for protection and safety consistent with day to day life;

(b) Establishment of an infrastructure to support continuing 'self-help protective actions' in the affected areas, such as by the provision of information and advice and by monitoring.

5.18. The conditions prevailing after the completion of remedial actions, if the regulatory body or other relevant authority has imposed no restrictions or controls, shall be considered to constitute the background conditions for any new facilities and activities or for habitation of the land.

Requirement 50: Public exposure due to radon indoors

The government shall provide information on levels of radon indoors and the associated health risks and, if appropriate, shall establish and implement an action plan for controlling public exposure due to radon indoors.

5.19. As part of its responsibilities as required in para. 5.3, the government shall ensure that:

(a) Information is gathered on activity concentrations of radon in dwellings and other buildings with high occupancy factors for members of the public[52] through appropriate means such as representative radon surveys;

(b) Relevant information on exposure due to radon and the associated health risks, including the increased risks relating to smoking, is provided to members of the public and other interested parties.

5.20. Where activity concentrations of radon that are of concern for public health are identified on the basis of the information gathered as required in para. 5.19(a), the government shall ensure that an action plan is established comprising coordinated actions to reduce radon levels for existing buildings and for future buildings, which includes[53]:

(a) Establishing an appropriate reference level for ^{222}Rn for dwellings and other buildings with high occupancy factors for members of the public with account taken of the prevailing social and economic circumstances that in general will not exceed an annual average activity concentration due to ^{222}Rn of 300 Bq/m^3 [54];

(b) Reducing activity concentrations of ^{222}Rn and consequent exposures to a level at which protection is optimized;

---

[52] Buildings with high occupancy factors for members of the public include kindergartens, schools and hospitals.

[53] Guidance on the preparation of an action plan for radon is provided, for example, in Ref. [6].

[54] On the assumption of an equilibrium factor for ^{222}Rn of 0.4 and an annual occupancy rate of 7000 hours, the value of activity concentration of 300 Bq/m^3 corresponds to an annual effective dose of the order of 10 mSv.

(c) Giving priority to reducing activity concentrations of ^{222}Rn in those situations for which such action is likely to be most effective[55];
(d) Including appropriate prevention and mitigation measures for ^{222}Rn exposure in building codes to prevent the ingress of radon and to facilitate possible remedial actions wherever necessary.

5.21. The government shall assign responsibility for:

(a) Establishing and implementing the action plan for controlling public exposure due to ^{222}Rn indoors;

(b) Determining the circumstances under which remedial action is to be mandatory or is to be voluntary, with account taken of legal requirements and of the prevailing social and economic circumstances.

Requirement 51: Exposure due to radionuclides in commodities

The regulatory body or other relevant authority shall establish reference levels for radionuclides in commodities.

5.22. The regulatory body or other relevant authority shall establish specific reference levels for exposure due to radionuclides in commodities such as construction material, food, feed and drinking water, each of which shall typically be expressed as, or based on, an annual effective dose to the representative person generally that does not exceed a value of about 1 mSv.

5.23. The regulatory body or other relevant authority shall consider the guideline levels for radionuclides contained in food traded internationally that could contain radioactive substances as a result of a nuclear or radiation emergency, as published by the Joint FAO/WHO Codex Alimentarius Commission [23]. The regulatory body or other relevant authority shall consider the guideline levels for radionuclides contained in drinking water that have been published by the WHO [24].

OCCUPATIONAL EXPOSURE

Scope

5.24. The requirements in respect of occupational exposure in existing exposure situations (paras 5.25–5.33) apply to any occupational exposure arising from the situations specified in para. 5.1.

Requirement 52: Exposure in workplaces

The regulatory body shall establish and enforce requirements for the protection of workers in existing exposure situations.

5.25. The requirements in respect of public exposure stated in paras 5.7–5.9 shall be applied for protection and safety for workers in existing exposure situations, other than in those specific situations identified in paras 5.26–5.33.

---

[55] Examples of giving priority to reducing activity concentrations of ^{222}Rn in those situations for which such action is likely to be most effective include (a) specifying levels of activity concentrations of ^{222}Rn in dwellings and other buildings with high occupancy factors at which protection can be considered optimized; (b) identifying radon prone areas; (c) identifying characteristics of buildings that are likely to give rise to elevated activity concentrations of ^{222}Rn ; and (d) identifying and requiring preventive measures for radon in future buildings that can be introduced at relatively low cost.

Remediation of areas with residual radioactive material

5.26. Employers shall ensure that the exposure of workers undertaking remedial actions is controlled in accordance with the relevant requirements for occupational exposure in planned exposure situations established in Section 3.

Exposure due to radon in workplaces

5.27. The regulatory body or other relevant authority shall establish a strategy for protection against exposure due to ^{222}Rn in workplaces, including the establishment of an appropriate reference level for ^{222}Rn. The reference level for ^{222}Rn shall be set at a value that does not exceed an annual average activity concentration of ^{222}Rn of 1000 Bq/m^3, with account taken of the prevailing social and economic circumstances.[56]

5.28. Employers shall ensure that activity concentrations of ^{222}Rn in workplaces are as low as reasonably achievable below the reference level established in accordance with para. 5.27, and shall ensure that protection is optimized.

5.29. If, despite all reasonable efforts by the employer to reduce radon levels, the activity concentration of ^{222}Rn in the workplace remains above the reference level established in accordance with para. 5.27, the relevant requirements for occupational exposure in planned exposure situations as stated in Section 3 shall apply.

Exposure of aircrew and space crew due to cosmic radiation

5.30. The regulatory body or other relevant authority shall determine whether assessment of the exposure of aircrew due to cosmic radiation is warranted.

5.31. Where such assessment is deemed to be warranted, the regulatory body or other relevant authority shall establish a framework which shall include a reference level of dose and a methodology for the assessment and recording of doses received by aircrew from occupational exposure to cosmic radiation.

5.32. In accordance with para. 5.31:

(a) Where the dose of aircrew members is likely to exceed the reference level, employers of aircrew:
   (i) shall assess doses and keep records;
   (ii) shall make records available to aircrew members;
(b) Employers:
   (i) shall inform female aircrew members of the risk to the embryo or fetus due to exposure to cosmic radiation and of the need for early notification of pregnancy;
   (ii) shall apply the requirements of para. 3.114 in respect of notification of pregnancy.

5.33. The regulatory body or other relevant authority shall establish, where appropriate, a framework for radiation protection that applies to individuals in space based activities that is appropriate for the exceptional conditions of space. While the requirements of these Standards in respect of dose limits do not apply to individuals in space based activities, all reasonable efforts shall be made to optimize protection by restricting the doses received by these individuals while not unduly limiting the extent of the activities that they undertake.

---

[56] On the assumption of an equilibrium factor for ^{222}Rn of 0.4 and an annual occupancy rate of 2000 hours, the value of activity concentration due to ^{222}Rn of 1000 Bq/m^3 corresponds to an annual effective dose of the order of 10 mSv.

Schedule I
EXEMPTION AND CLEARANCE

CRITERIA FOR EXEMPTION

I-1. The general criteria for exemption are that:

(a) Radiation risks arising from the practice or a source within a practice are sufficiently low as not to warrant regulatory control, with no appreciable likelihood of situations that could lead to a failure to meet the general criterion for exemption; or
(b) Regulatory control of the practice or the source would yield no net benefit, in that no reasonable control measures would achieve a worthwhile return in terms of reduction of individual doses or of health risks.

I-2. A practice or a source within a practice may be exempted under the terms of para. I-1(a) without further consideration provided that under all reasonably foreseeable circumstances the effective dose expected to be incurred by any member of the public (normally evaluated on the basis of a safety assessment) owing to the exempted practice or the exempted source within the practice is of the order of 10 μSv or less in a year. To take into account low probability scenarios, a different criterion could be used, namely that the effective dose expected to be incurred by any member of the public for such low probability scenarios does not exceed 1 mSv in a year.

I-3. Under the criteria set out in paras I-1 and I-2, the following sources within justified practices are automatically exempted without further consideration from the requirements of these Standards, including requirements for notification, registration or licensing:

(a) Radioactive material in a moderate amount[57] for which either the total activity of an individual radionuclide present on the premises at any one time or the activity concentration as used in the practice does not exceed the applicable exemption level given in Table I-1 of Schedule I;[58]
(b) Radioactive material in bulk amount[57] for which the activity concentration of a given radionuclide of artificial origin used in the practice does not exceed the relevant value given in Table I-2 of Schedule I[58];
(c) Radiation generators of a type approved by the regulatory body, or in the form of an electronic tube, such as a cathode ray tube for the display of visual images, provided that:

    (i) They do not in normal operating conditions cause an ambient dose equivalent rate or a directional dose equivalent rate, as appropriate, exceeding 1 μSv/h at a distance of 0.1 m from any accessible surface of the equipment; or
    (ii) The maximum energy of the radiation generated is no greater than 5 keV.

---

[57] The exemption values (activity concentrations) presented in Table I-1 have been calculated on the basis of scenarios involving a moderate amount of material: "The calculated values apply to practices involving small scale usage of activity where the quantities involved are at the most of the order of a tonne" (see Ref. [25]). The regulatory body will need to establish the amounts for which the concentration values in Table I-1 may be applied, bearing in mind that for many radionuclides, in particular those for which there is no corresponding value given in Table I-2, a restriction on the amount is not meaningful.

[58] The exemption levels set out in Table I-1 and the exemption and clearance levels set out in Table I-2 of Schedule I are subject to the following considerations: (a) They were derived using a conservative model based on (i) the criteria of paras I-2 and I-11 respectively and (ii) a series of limiting (bounding) scenarios for use and disposal (see Ref. [25] in the case of Table I-1 and Ref. [26] in the case of Table I-2). (b) If there is more than one radionuclide, the derived exemption level or derived clearance level for the mixture is determined as specified in paras I-7 and I-14.

I-4. For radionuclides of natural origin, exemption of bulk amounts of material is necessarily considered on a case by case basis[59] by using a dose criterion of the order of 1 mSv in a year, commensurate with typical doses due to natural background levels of radiation.

I-5. The IAEA Regulations for the Safe Transport of Radioactive Material [12] (the Transport Regulations) do not apply to exempt material or exempt consignments; that is, they do not apply to material in transport for which the activity concentration of the material (for exempt material) or the total activity of radionuclides in the consignment (for an exempt consignment) does not exceed the relevant 'basic radionuclide value' for exemption given in the Transport Regulations[60]. Usually, such basic radionuclide values are numerically equal to the corresponding exempt activity concentrations or exempt activities given in Table I-1 of Schedule I.

I-6. Exemptions may be granted subject to conditions specified by the regulatory body, such as conditions relating to the physical or chemical form of the radioactive material, and to its use or the means of its disposal. In particular, such an exemption may be granted for equipment containing radioactive material that is not otherwise exempted under para. I-3(a) provided that:

(a) The equipment containing radioactive material is of a type approved by the regulatory body;

(b) The radioactive material:
   (i) Is in the form of a sealed source that effectively prevents any contact with the radioactive material and prevents its leakage; or
   (ii) Is in the form of an unsealed source in a small amount such as sources used for radioimmunoassay;

(c) In normal operating conditions the equipment does not cause an ambient dose equivalent rate or a directional dose equivalent rate, as appropriate, exceeding 1 μSv/h at a distance of 0.1 m from any accessible surface of the equipment;

(d) Necessary conditions for disposal of the equipment have been specified by the regulatory body.

I-7. For exemption of radioactive material containing more than one radionuclide, on the basis of the levels given in Table I-1 and Table I-2, the condition for exemption is that the sum of the individual radionuclide activities or activity concentrations, as appropriate, is less than the derived exemption level for the mixture (X_m), determined as follows:

$$X_m = \frac{1}{\sum_{i=1}^{n} \frac{f(i)}{X(i)}}$$

where *f(i)* is the fraction of activity or activity concentration, as appropriate, of radionuclide *i* in the mixture, *X(i)* is the applicable level for radionuclide *i* as given in Table I-1 or Table I-2, and *n* is the number of radionuclides present.

I-8. Radioactive material arising from authorized discharges is exempted from any requirements for notification, registration or licensing unless otherwise specified by the regulatory body.

---

[59] Material containing radionuclides of natural origin at an activity concentration of less than 1 Bq/g for any radionuclide in the uranium and thorium decay chains and of less than 10 Bq/g for ^{40}K is outside the scope of planned exposure situations (para. 3.4(a)); hence the concept of exemption does not apply for such material.

[60] For the purposes of material in transport, 'exemption' means exemption from the requirements of the IAEA Transport Regulations [12].

I-9. The values provided in Table I-1 and Table I-2 are not intended to be applied to the control of discharges or the control of radioactive residues in the environment.

CRITERIA FOR CLEARANCE

I-10. The general criteria for clearance are that:

(a) Radiation risks arising from the cleared material are sufficiently low as not to warrant regulatory control, with no appreciable likelihood of occurrence for scenarios that could lead to a failure to meet the general criterion for clearance; or
(b) Continued regulatory control of the material would yield no net benefit, in that no reasonable control measures would achieve a worthwhile return in terms of reduction of individual doses or of health risks.

I-11. Material may be cleared under the terms of para. I-10(a) without further consideration provided that in all reasonably foreseeable situations the effective dose expected to be incurred by any member of the public due to the cleared material is of the order of 10 μSv or less in a year. To take into account low probability scenarios, a different criterion can be used, namely that the effective dose expected to be incurred by any member of the public for such low probability scenarios does not exceed 1 mSv in a year.

I-12. Radioactive material within a notified practice or an authorized practice may be cleared without further consideration provided that:

(a) The activity concentration of an individual radionuclide of artificial origin in solid form does not exceed the relevant level given in Table I-2 of Schedule I[58]; or
(b) The activity concentrations of radionuclides of natural origin do not exceed the relevant level given in Table I-3 of Schedule I[61]; or
(c) For radionuclides of natural origin in residues that might be recycled into construction materials[62] or the disposal of which is liable to cause the contamination of drinking water supplies, the activity concentration in the residues does not exceed specific values derived so as to meet a dose criterion of the order of 1 mSv in a year, commensurate with typical doses due to natural background levels of radiation.

I-13. Clearance may be granted by the regulatory body for specific situations, on the basis of the criteria of paras I-10 and I-11, with account taken of the physical or chemical form of the radioactive material, and its use or the means of its disposal[63]. Such clearance levels may be specified in terms of activity concentration per unit mass or per unit surface area.

I-14. For clearance of radioactive material containing more than one radionuclide of artificial origin, on the basis of the levels given in Table I-2, the condition for clearance is that the sum of the individual radionuclide activity concentrations is less than the derived clearance level for the mixture (X_m), determined as follows:

---

[61] These values of activity concentration may also be applied for the clearance of materials arising from practices subject to the clearance criteria given in para. I-11, pending the establishment of radionuclide specific values for the radionuclides of natural origin given in Table I-3.

[62] Regulatory control of construction materials is addressed in Section 5 as an existing exposure situation.

[63] For example, specific clearance levels may be developed for metals, rubble from buildings and waste for disposal in landfill sites.

$$X_m = \frac{1}{\sum_{i=1}^{n} \frac{f(i)}{X(i)}}$$

where *f(i)* is the fraction of activity concentration of radionuclide *i* in the mixture, *X(i)* is the applicable level for radionuclide *i* as given in Table I-2, and *n* is the number of radionuclides present.

I-15. For clearance of bulk quantities of material containing a mixture of radionuclides of natural origin and radionuclides of artificial origin, the conditions given in paras I-12(b) and I-14 both have to be satisfied.

TABLE I-1: LEVELS FOR EXEMPTION OF MODERATE AMOUNTS OF MATERIAL WITHOUT FURTHER CONSIDERATION: EXEMPT ACTIVITY CONCENTRATIONS AND EXEMPT ACTIVITIES OF RADIONUCLIDES *(see footnotes 57 and 58)*

Radionuclide	Activity concentration (Bq/g)	Activity (Bq)
H-3	1×10^6	1×10^9
Be-7	1×10^3	1×10^7
Be-10	1×10^4	1×10^6
C-11	1×10^1	1×10^6
C-14	1×10^4	1×10^7
N-13	1×10^2	1×10^9
Ne-19	1×10^2	1×10^9
O-15	1×10^2	1×10^9
F-18	1×10^1	1×10^6
Na-22	1×10^1	1×10^6
Na-24	1×10^1	1×10^5
Mg-28	1×10^1	1×10^5
Al-26	1×10^1	1×10^5
Si-31	1×10^3	1×10^6
Si-32	1×10^3	1×10^6
P-32	1×10^3	1×10^5
P-33	1×10^5	1×10^8
S-35	1×10^5	1×10^8
Cl-36	1×10^4	1×10^6
Cl-38	1×10^1	1×10^5
Cl-39	1×10^1	1×10^5
Ar-37	1×10^6	1×10^8
Ar-39	1×10^7	1×10^4
Ar-41	1×10^2	1×10^9
K-40	1×10^2	1×10^6
K-42	1×10^2	1×10^6
K-43	1×10^1	1×10^6
K-44	1×10^1	1×10^5
K-45	1×10^1	1×10^5
Ca-41	1×10^5	1×10^7
Ca-45	1×10^4	1×10^7
Ca-47	1×10^1	1×10^6
Sc-43	1×10^1	1×10^6
Sc-44	1×10^1	1×10^5
Sc-45	1×10^2	1×10^7
Sc-46	1×10^1	1×10^6
Sc-47	1×10^2	1×10^6

Radionuclide	Activity concentration (Bq/g)	Activity (Bq)
Sc-48	1×10^1	1×10^5
Sc-49	1×10^3	1×10^5
Ti-44	1×10^1	1×10^5
Ti-45	1×10^1	1×10^6
V-47	1×10^1	1×10^5
V-48	1×10^1	1×10^5
V-49	1×10^4	1×10^7
Cr-48	1×10^2	1×10^6
Cr-49	1×10^1	1×10^6
Cr-51	1×10^3	1×10^7
Mn-51	1×10^1	1×10^5
Mn-52	1×10^1	1×10^5
Mn-52m	1×10^1	1×10^5
Mn-53	1×10^4	1×10^9
Mn-54	1×10^1	1×10^6
Mn-56	1×10^1	1×10^5
Fe-52	1×10^1	1×10^6
Fe-55	1×10^4	1×10^6
Fe-59	1×10^1	1×10^6
Fe-60	1×10^2	1×10^5
Co-55	1×10^1	1×10^6
Co-56	1×10^1	1×10^5
Co-57	1×10^2	1×10^6
Co-58	1×10^1	1×10^6
Co-58m	1×10^4	1×10^7
Co-60	1×10^1	1×10^5
Co-60m	1×10^3	1×10^6
Co-61	1×10^2	1×10^6
Co-62m	1×10^1	1×10^5
Ni-56	1×10^1	1×10^6
Ni-57	1×10^1	1×10^6
Ni-59	1×10^4	1×10^8
Ni-63	1×10^5	1×10^8
Ni-65	1×10^1	1×10^6
Ni-66	1×10^4	1×10^7
Cu-60	1×10^1	1×10^5
Cu-61	1×10^1	1×10^6

Radionuclide	Activity concentration (Bq/g)	Activity (Bq)
Cu-64	1×10^{2}	1×10^{6}
Cu-67	1×10^{2}	1×10^{6}
Zn-62	1×10^{2}	1×10^{6}
Zn-63	1×10^{1}	1×10^{5}
Zn-65	1×10^{1}	1×10^{6}
Zn-69	1×10^{4}	1×10^{6}
Zn-69m	1×10^{2}	1×10^{6}
Zn-71m	1×10^{1}	1×10^{6}
Zn-72	1×10^{2}	1×10^{6}
Ga-65	1×10^{1}	1×10^{5}
Ga-66	1×10^{1}	1×10^{5}
Ga-67	1×10^{2}	1×10^{6}
Ga-68	1×10^{1}	1×10^{5}
Ga-70	1×10^{2}	1×10^{6}
Ga-72	1×10^{1}	1×10^{5}
Ga-73	1×10^{2}	1×10^{6}
Ge-66	1×10^{1}	1×10^{6}
Ge-67	1×10^{1}	1×10^{5}
Ge-68[a]	1×10^{1}	1×10^{5}
Ge-69	1×10^{1}	1×10^{6}
Ge-71	1×10^{4}	1×10^{8}
Ge-75	1×10^{3}	1×10^{6}
Ge-77	1×10^{1}	1×10^{5}
Ge-78	1×10^{2}	1×10^{6}
As-69	1×10^{1}	1×10^{5}
As-70	1×10^{1}	1×10^{5}
As-71	1×10^{1}	1×10^{6}
As-72	1×10^{1}	1×10^{5}
As-73	1×10^{3}	1×10^{7}
As-74	1×10^{1}	1×10^{6}
As-76	1×10^{2}	1×10^{5}
As-77	1×10^{3}	1×10^{6}
As-78	1×10^{1}	1×10^{5}
Se-70	1×10^{1}	1×10^{6}
Se-73	1×10^{1}	1×10^{6}
Se-73m	1×10^{2}	1×10^{6}
Se-75	1×10^{2}	1×10^{6}
Se-79	1×10^{4}	1×10^{7}
Se-81	1×10^{3}	1×10^{6}
Se-81m	1×10^{3}	1×10^{7}
Se-83	1×10^{1}	1×10^{5}
Br-74	1×10^{1}	1×10^{5}
Br-74m	1×10^{1}	1×10^{5}
Br-75	1×10^{1}	1×10^{6}
Br-76	1×10^{1}	1×10^{5}
Br-77	1×10^{2}	1×10^{6}
Br-80	1×10^{2}	1×10^{5}
Br-80m	1×10^{3}	1×10^{7}
Br-82	1×10^{1}	1×10^{6}
Br-83	1×10^{3}	1×10^{6}
Br-84	1×10^{1}	1×10^{5}
Kr-74	1×10^{2}	1×10^{9}
Kr-76	1×10^{2}	1×10^{9}
Kr-77	1×10^{2}	1×10^{9}
Kr-79	1×10^{3}	1×10^{5}
Kr-81	1×10^{4}	1×10^{7}
Kr-81m	1×10^{3}	1×10^{10}
Kr-83m	1×10^{5}	1×10^{12}
Kr-85	1×10^{5}	1×10^{4}
Kr-85m	1×10^{3}	1×10^{10}
Kr-87	1×10^{2}	1×10^{9}
Kr-88	1×10^{2}	1×10^{9}
Rb-79	1×10^{1}	1×10^{5}
Rb-81	1×10^{1}	1×10^{6}
Rb-81m	1×10^{3}	1×10^{7}
Rb-82m	1×10^{1}	1×10^{6}
Rb-83[a]	1×10^{2}	1×10^{6}
Rb-84	1×10^{1}	1×10^{6}
Rb-86	1×10^{2}	1×10^{5}
Rb-87	1×10^{3}	1×10^{7}
Rb-88	1×10^{2}	1×10^{5}
Rb-89	1×10^{2}	1×10^{5}
Sr-80	1×10^{3}	1×10^{7}
Sr-81	1×10^{1}	1×10^{5}
Sr-82[a]	1×10^{1}	1×10^{5}
Sr-83	1×10^{1}	1×10^{6}
Sr-85	1×10^{2}	1×10^{6}
Sr-85m	1×10^{2}	1×10^{7}
Sr-87m	1×10^{2}	1×10^{6}
Sr-89	1×10^{3}	1×10^{6}
Sr-90[a]	1×10^{2}	1×10^{4}
Sr-91	1×10^{1}	1×10^{5}

Radionuclide	Activity concentration (Bq/g)	Activity (Bq)
Sr-92	1×10^1	1×10^6
Y-86	1×10^1	1×10^5
Y-86m	1×10^2	1×10^7
Y-87[a]	1×10^1	1×10^6
Y-88	1×10^1	1×10^6
Y-90	1×10^3	1×10^5
Y-90m	1×10^1	1×10^6
Y-91	1×10^3	1×10^6
Y-91m	1×10^2	1×10^6
Y-92	1×10^2	1×10^5
Y-93	1×10^2	1×10^5
Y-94	1×10^1	1×10^5
Y-95	1×10^1	1×10^5
Zr-86	1×10^2	1×10^7
Zr-88	1×10^2	1×10^6
Zr-89	1×10^1	1×10^6
Zr-93[a]	1×10^3	1×10^7
Zr-95	1×10^1	1×10^6
Zr-97[a]	1×10^1	1×10^5
Nb-88	1×10^1	1×10^5
Nb-89 (2.03 h)	1×10^1	1×10^5
Nb-89m (1.10 h)	1×10^1	1×10^5
Nb-90	1×10^1	1×10^5
Nb-93m	1×10^4	1×10^7
Nb-94	1×10^1	1×10^6
Nb-95	1×10^1	1×10^6
Nb-95m	1×10^2	1×10^7
Nb-96	1×10^1	1×10^5
Nb-97	1×10^1	1×10^6
Nb-98	1×10^1	1×10^5
Mo-90	1×10^1	1×10^6
Mo-93	1×10^3	1×10^8
Mo-93m	1×10^1	1×10^6
Mo-99	1×10^2	1×10^6
Mo-101	1×10^1	1×10^6
Tc-93	1×10^1	1×10^6
Tc-93m	1×10^1	1×10^6
Tc-94	1×10^1	1×10^6
Tc-94m	1×10^1	1×10^5
Tc-95	1×10^1	1×10^6
Tc-95m	1×10^1	1×10^6
Tc-96	1×10^1	1×10^6
Tc-96m	1×10^3	1×10^7
Tc-97	1×10^3	1×10^8
Tc-97m	1×10^3	1×10^7
Tc-98	1×10^1	1×10^6
Tc-99	1×10^4	1×10^7
Tc-99m	1×10^2	1×10^7
Tc-101	1×10^2	1×10^6
Tc-104	1×10^1	1×10^5
Ru-94	1×10^2	1×10^6
Ru-97	1×10^2	1×10^7
Ru-103	1×10^2	1×10^6
Ru-105	1×10^1	1×10^6
Ru-106[a]	1×10^2	1×10^5
Rh-99	1×10^1	1×10^6
Rh-99m	1×10^1	1×10^6
Rh-100	1×10^1	1×10^6
Rh-101	1×10^2	1×10^7
Rh-101m	1×10^2	1×10^7
Rh-102	1×10^1	1×10^6
Rh-102m	1×10^2	1×10^6
Rh-103m	1×10^4	1×10^8
Rh-105	1×10^2	1×10^7
Rh-106m	1×10^1	1×10^5
Rh-107	1×10^2	1×10^6
Pd-100	1×10^2	1×10^7
Pd-101	1×10^2	1×10^6
Pd-103	1×10^3	1×10^8
Pd-107	1×10^5	1×10^8
Pd-109	1×10^3	1×10^6
Ag-102	1×10^1	1×10^5
Ag-103	1×10^1	1×10^6
Ag-104	1×10^1	1×10^6
Ag-104m	1×10^1	1×10^6
Ag-105	1×10^2	1×10^6
Ag-106	1×10^1	1×10^6
Ag-106m	1×10^1	1×10^6
Ag-108m	1×10^1	1×10^6
Ag-110m	1×10^1	1×10^6
Ag-111	1×10^3	1×10^6
Ag-112	1×10^1	1×10^5

Radionuclide	Activity concentration (Bq/g)	Activity (Bq)	Radionuclide	Activity concentration (Bq/g)	Activity (Bq)
Ag-115	1×10^1	1×10^5	Sb-118m	1×10^1	1×10^6
Cd-104	1×10^2	1×10^7	Sb-119	1×10^3	1×10^7
Cd-107	1×10^3	1×10^7	Sb-120 (5.76d)	1×10^1	1×10^6
Cd-109	1×10^4	1×10^6	Sb-120 (15.89m)	1×10^2	1×10^6
Cd-113	1×10^3	1×10^6	Sb-122	1×10^2	1×10^4
Cd-113m	1×10^3	1×10^6	Sb-124	1×10^1	1×10^6
Cd-115	1×10^2	1×10^6	Sb-124m	1×10^2	1×10^6
Cd-115m	1×10^3	1×10^6	Sb-125	1×10^2	1×10^6
Cd-117	1×10^1	1×10^6	Sb-126	1×10^1	1×10^5
Cd-117m	1×10^1	1×10^6	Sb-126m	1×10^1	1×10^5
In-109	1×10^1	1×10^6	Sb-127	1×10^1	1×10^6
In-110 (4.9h)	1×10^1	1×10^6	Sb-128(9.01h)	1×10^1	1×10^5
In-110 (69.1m)	1×10^1	1×10^5	Sb-128 (10.4m)	1×10^1	1×10^5
In-111	1×10^2	1×10^6	Sb-129	1×10^1	1×10^6
In-112	1×10^2	1×10^6	Sb-130	1×10^1	1×10^5
In-113m	1×10^2	1×10^6	Sb-131	1×10^1	1×10^6
In-114	1×10^3	1×10^5	Te-116	1×10^2	1×10^7
In-114m	1×10^2	1×10^6	Te-121	1×10^1	1×10^6
In-115	1×10^3	1×10^5	Te-121m	1×10^2	1×10^6
In-115m	1×10^2	1×10^6	Te-123	1×10^3	1×10^6
In-116m	1×10^1	1×10^5	Te-123m	1×10^2	1×10^7
In-117	1×10^1	1×10^6	Te-125m	1×10^3	1×10^7
In-117m	1×10^2	1×10^6	Te-127	1×10^3	1×10^6
In-119m	1×10^2	1×10^5	Te-127m	1×10^3	1×10^7
Sn-110	1×10^2	1×10^7	Te-129	1×10^2	1×10^6
Sn-111	1×10^2	1×10^6	Te-129m	1×10^3	1×10^6
Sn-113	1×10^3	1×10^7	Te-131	1×10^2	1×10^5
Sn-117m	1×10^2	1×10^6	Te-131m	1×10^1	1×10^6
Sn-119m	1×10^3	1×10^7	Te-132	1×10^2	1×10^7
Sn-121	1×10^5	1×10^7	Te-133	1×10^1	1×10^5
Sn-121m[a]	1×10^3	1×10^7	Te-133m	1×10^1	1×10^5
Sn-123	1×10^3	1×10^6	Te-134	1×10^1	1×10^6
Sn-123m	1×10^2	1×10^6	I-120	1×10^1	1×10^5
Sn-125	1×10^2	1×10^5	I-120m	1×10^1	1×10^5
Sn-126[a]	1×10^1	1×10^5	I-121	1×10^2	1×10^6
Sn-127	1×10^1	1×10^6	I-123	1×10^2	1×10^7
Sn-128	1×10^1	1×10^6	I-124	1×10^1	1×10^6
Sb-115	1×10^1	1×10^6	I-125	1×10^3	1×10^6
Sb-116	1×10^1	1×10^6	I-126	1×10^2	1×10^6
Sb-116m	1×10^1	1×10^5	I-128	1×10^2	1×10^5
Sb-117	1×10^2	1×10^7	I-129	1×10^2	1×10^5

Radionuclide	Activity concentration (Bq/g)	Activity (Bq)
I-130	1×10^{1}	1×10^{6}
I-131	1×10^{2}	1×10^{6}
I-132	1×10^{1}	1×10^{5}
I-132m	1×10^{2}	1×10^{6}
I-133	1×10^{1}	1×10^{6}
I-134	1×10^{1}	1×10^{5}
I-135	1×10^{1}	1×10^{6}
Xe-120	1×10^{2}	1×10^{9}
Xe-121	1×10^{2}	1×10^{9}
Xe-122[a]	1×10^{2}	1×10^{9}
Xe-123	1×10^{2}	1×10^{9}
Xe-125	1×10^{3}	1×10^{9}
Xe-127	1×10^{3}	1×10^{5}
Xe-129m	1×10^{3}	1×10^{4}
Xe-131m	1×10^{4}	1×10^{4}
Xe-133m	1×10^{3}	1×10^{4}
Xe-133	1×10^{3}	1×10^{4}
Xe-135	1×10^{3}	1×10^{10}
Xe-135m	1×10^{2}	1×10^{9}
Xe-138	1×10^{2}	1×10^{9}
Cs-125	1×10^{1}	1×10^{4}
Cs-127	1×10^{2}	1×10^{5}
Cs-129	1×10^{2}	1×10^{5}
Cs-130	1×10^{2}	1×10^{6}
Cs-131	1×10^{3}	1×10^{6}
Cs-132	1×10^{1}	1×10^{5}
Cs-134m	1×10^{3}	1×10^{5}
Cs-134	1×10^{1}	1×10^{4}
Cs-135	1×10^{4}	1×10^{7}
Cs-135m	1×10^{1}	1×10^{6}
Cs-136	1×10^{1}	1×10^{5}
Cs-137[a]	1×10^{1}	1×10^{4}
Cs-138	1×10^{1}	1×10^{4}
Ba-126	1×10^{2}	1×10^{7}
Ba-128	1×10^{2}	1×10^{7}
Ba-131	1×10^{2}	1×10^{6}
Ba-131m	1×10^{2}	1×10^{7}
Ba-133	1×10^{2}	1×10^{6}
Ba-133m	1×10^{2}	1×10^{6}
Ba-135m	1×10^{2}	1×10^{6}
Ba-137m	1×10^{1}	1×10^{6}

Radionuclide	Activity concentration (Bq/g)	Activity (Bq)
Ba-139	1×10^{2}	1×10^{5}
Ba-140[a]	1×10^{1}	1×10^{5}
Ba-141	1×10^{2}	1×10^{5}
Ba-142	1×10^{2}	1×10^{6}
La-131	1×10^{1}	1×10^{6}
La-132	1×10^{1}	1×10^{6}
La-135	1×10^{3}	1×10^{7}
La-137	1×10^{3}	1×10^{7}
La-138	1×10^{1}	1×10^{6}
La-140	1×10^{1}	1×10^{5}
La-141	1×10^{2}	1×10^{5}
La-142	1×10^{1}	1×10^{5}
La-143	1×10^{2}	1×10^{5}
Ce-134	1×10^{3}	1×10^{7}
Ce-135	1×10^{1}	1×10^{6}
Ce-137	1×10^{3}	1×10^{7}
Ce-137m	1×10^{3}	1×10^{6}
Ce-139	1×10^{2}	1×10^{6}
Ce-141	1×10^{2}	1×10^{7}
Ce-143	1×10^{2}	1×10^{6}
Ce-144[a]	1×10^{2}	1×10^{5}
Pr-136	1×10^{1}	1×10^{5}
Pr-137	1×10^{2}	1×10^{6}
Pr-138m	1×10^{1}	1×10^{6}
Pr-139	1×10^{2}	1×10^{7}
Pr-142	1×10^{2}	1×10^{5}
Pr-142m	1×10^{7}	1×10^{9}
Pr-143	1×10^{4}	1×10^{6}
Pr-144	1×10^{2}	1×10^{5}
Pr-145	1×10^{3}	1×10^{5}
Pr-147	1×10^{1}	1×10^{5}
Nd-136	1×10^{2}	1×10^{6}
Nd-138	1×10^{3}	1×10^{7}
Nd-139	1×10^{2}	1×10^{6}
Nd-139m	1×10^{1}	1×10^{6}
Nd-141	1×10^{2}	1×10^{7}
Nd-147	1×10^{2}	1×10^{6}
Nd-149	1×10^{2}	1×10^{6}
Nd-151	1×10^{1}	1×10^{5}
Pm-141	1×10^{1}	1×10^{5}
Pm-143	1×10^{2}	1×10^{6}

Radionuclide	Activity concentration (Bq/g)	Activity (Bq)
Pm-144	1×10^1	1×10^6
Pm-145	1×10^3	1×10^7
Pm-146	1×10^1	1×10^6
Pm-147	1×10^4	1×10^7
Pm-148	1×10^1	1×10^5
Pm-148m	1×10^1	1×10^6
Pm-149	1×10^3	1×10^6
Pm-150	1×10^1	1×10^5
Pm-151	1×10^2	1×10^6
Sm-141	1×10^1	1×10^5
Sm-141m	1×10^1	1×10^6
Sm-142	1×10^2	1×10^7
Sm-145	1×10^2	1×10^7
Sm-146	1×10^1	1×10^5
Sm-147	1×10^1	1×10^4
Sm-151	1×10^4	1×10^8
Sm-153	1×10^2	1×10^6
Sm-155	1×10^2	1×10^6
Sm-156	1×10^2	1×10^6
Eu-145	1×10^1	1×10^6
Eu-146	1×10^1	1×10^6
Eu-147	1×10^2	1×10^6
Eu-148	1×10^1	1×10^6
Eu-149	1×10^2	1×10^7
Eu-150 (34.2a)	1×10^1	1×10^6
Eu-150 (12.6h)	1×10^3	1×10^6
Eu-152	1×10^1	1×10^6
Eu-152m	1×10^2	1×10^6
Eu-154	1×10^1	1×10^6
Eu-155	1×10^2	1×10^7
Eu-156	1×10^1	1×10^6
Eu157	1×10^2	1×10^6
Eu-158	1×10^1	1×10^5
Gd-145	1×10^1	1×10^5
Gd-146[a]	1×10^1	1×10^6
Gd-147	1×10^1	1×10^6
Gd-148	1×10^1	1×10^4
Gd-149	1×10^2	1×10^6
Gd-151	1×10^2	1×10^7
Gd-152	1×10^1	1×10^4
Gd-153	1×10^2	1×10^7
Gd-159	1×10^3	1×10^6
Tb-147	1×10^1	1×10^6
Tb-149	1×10^1	1×10^6
Tb-150	1×10^1	1×10^6
Tb-151	1×10^1	1×10^6
Tb-153	1×10^2	1×10^7
Tb-154	1×10^1	1×10^6
Tb-155	1×10^2	1×10^7
Tb-156	1×10^1	1×10^6
Tb-156m (24.4h)	1×10^3	1×10^7
Tb-156m (5h)	1×10^4	1×10^7
Tb-157	1×10^4	1×10^7
Tb-158	1×10^1	1×10^6
Tb-160	1×10^1	1×10^6
Tb-161	1×10^3	1×10^6
Dy-155	1×10^1	1×10^6
Dy-157	1×10^2	1×10^6
Dy-159	1×10^3	1×10^7
Dy-165	1×10^3	1×10^6
Dy-166	1×10^3	1×10^6
Ho-155	1×10^2	1×10^6
Ho-157	1×10^2	1×10^6
Ho-159	1×10^2	1×10^6
Ho-161	1×10^2	1×10^7
Ho-162	1×10^2	1×10^7
Ho-162m	1×10^1	1×10^6
Ho-164	1×10^3	1×10^6
Ho-164m	1×10^3	1×10^7
Ho-166	1×10^3	1×10^5
Ho-166m	1×10^1	1×10^6
Ho-167	1×10^2	1×10^6
Er-161	1×10^1	1×10^6
Er-165	1×10^3	1×10^7
Er-169	1×10^4	1×10^7
Er-171	1×10^2	1×10^6
Er-172	1×10^2	1×10^6
Tm-162	1×10^1	1×10^6
Tm-166	1×10^1	1×10^6
Tm-167	1×10^2	1×10^6
Tm-170	1×10^3	1×10^6
Tm-171	1×10^4	1×10^8

Radionuclide	Activity concentration (Bq/g)	Activity (Bq)
Tm-172	1×10^2	1×10^6
Tm-173	1×10^2	1×10^6
Tm-175	1×10^1	1×10^6
Yb-162	1×10^2	1×10^7
Yb-166	1×10^2	1×10^7
Yb-167	1×10^2	1×10^6
Yb-169	1×10^2	1×10^7
Yb-175	1×10^3	1×10^7
Yb-177	1×10^2	1×10^6
Yb-178	1×10^3	1×10^6
Lu-169	1×10^1	1×10^6
Lu-170	1×10^1	1×10^6
Lu-171	1×10^1	1×10^6
Lu-172	1×10^1	1×10^6
Lu-173	1×10^2	1×10^7
Lu-174	1×10^2	1×10^7
Lu-174m	1×10^2	1×10^7
Lu-176	1×10^2	1×10^6
Lu-176m	1×10^3	1×10^6
Lu-177	1×10^3	1×10^7
Lu-177m	1×10^1	1×10^6
Lu-178	1×10^2	1×10^5
Lu-178m	1×10^1	1×10^5
Lu-179	1×10^3	1×10^6
Hf-170	1×10^2	1×10^6
Hf-172[a]	1×10^1	1×10^6
Hf-173	1×10^2	1×10^6
Hf-175	1×10^2	1×10^6
Hf-177m	1×10^1	1×10^5
Hf-178m	1×10^1	1×10^6
Hf-179m	1×10^1	1×10^6
Hf-180m	1×10^1	1×10^6
Hf-181	1×10^1	1×10^6
Hf-182	1×10^2	1×10^6
Hf-182m	1×10^1	1×10^6
Hf-183	1×10^1	1×10^6
Hf-184	1×10^2	1×10^6
Ta-172	1×10^1	1×10^6
Ta -173	1×10^1	1×10^6
Ta-174	1×10^1	1×10^6
Ta-175	1×10^1	1×10^6
Ta-176	1×10^1	1×10^6
Ta-177	1×10^2	1×10^7
Ta-178	1×10^1	1×10^6
Ta-179	1×10^3	1×10^7
Ta-180	1×10^1	1×10^6
Ta-180m	1×10^3	1×10^7
Ta-182	1×10^1	1×10^4
Ta-182m	1×10^2	1×10^6
Ta-183	1×10^2	1×10^6
Ta-184	1×10^1	1×10^6
Ta-185	1×10^2	1×10^5
Ta-186	1×10^1	1×10^5
W-176	1×10^2	1×10^6
W-177	1×10^1	1×10^6
W-178[a]	1×10^1	1×10^6
W-179	1×10^2	1×10^7
W-181	1×10^3	1×10^7
W-185	1×10^4	1×10^7
W-187	1×10^2	1×10^6
W-188[a]	1×10^2	1×10^5
Re-177	1×10^1	1×10^6
Re-178	1×10^1	1×10^6
Re-181	1×10^1	1×10^6
Re-182 (64h)	1×10^1	1×10^6
Re-182 (12.7h)	1×10^1	1×10^6
Re-184	1×10^1	1×10^6
Re-184m	1×10^2	1×10^6
Re-186	1×10^3	1×10^6
Re-186m	1×10^3	1×10^7
Re-187	1×10^6	1×10^9
Re-188	1×10^2	1×10^5
Re-188m	1×10^2	1×10^7
Re-189[a]	1×10^2	1×10^6
Os-180	1×10^2	1×10^7
Os-181	1×10^1	1×10^6
Os-182	1×10^2	1×10^6
Os-185	1×10^1	1×10^6
Os-189m	1×10^4	1×10^7
Os-191	1×10^2	1×10^7
Os-191m	1×10^3	1×10^7
Os-193	1×10^2	1×10^6

Radionuclide	Activity concentration (Bq/g)	Activity (Bq)
Os-194[a]	1×10^2	1×10^5
Ir-182	1×10^1	1×10^5
Ir-184	1×10^1	1×10^6
Ir-185	1×10^1	1×10^6
Ir-186 (15.8h)	1×10^1	1×10^6
Ir-186 (1.75h)	1×10^1	1×10^6
Ir-187	1×10^2	1×10^6
Ir-188	1×10^1	1×10^6
Ir-189[a]	1×10^2	1×10^7
Ir-190	1×10^1	1×10^6
Ir-190m (3.1h)	1×10^1	1×10^6
Ir-190m (1.2h)	1×10^4	1×10^7
Ir-192	1×10^1	1×10^4
Ir-192m	1×10^2	1×10^7
Ir-193m	1×10^4	1×10^7
Ir-194	1×10^2	1×10^5
Ir-194m	1×10^1	1×10^6
Ir-195	1×10^2	1×10^6
Ir-195m	1×10^2	1×10^6
Pt-186	1×10^1	1×10^6
Pt-188[a]	1×10^1	1×10^6
Pt-189	1×10^2	1×10^6
Pt-191	1×10^2	1×10^6
Pt-193	1×10^4	1×10^7
Pt-193m	1×10^3	1×10^7
Pt-195m	1×10^2	1×10^6
Pt-197	1×10^3	1×10^6
Pt-197m	1×10^2	1×10^6
Pt-199	1×10^2	1×10^6
Pt-200	1×10^2	1×10^6
Au-193	1×10^2	1×10^7
Au-194	1×10^1	1×10^6
Au-195	1×10^2	1×10^7
Au-198	1×10^2	1×10^6
Au-198m	1×10^1	1×10^6
Au-199	1×10^2	1×10^6
Au-200	1×10^2	1×10^5
Au-200m	1×10^1	1×10^6
Au-201	1×10^2	1×10^6
Hg-193	1×10^2	1×10^6
Hg-193m	1×10^1	1×10^6

Radionuclide	Activity concentration (Bq/g)	Activity (Bq)
Hg-194[a]	1×10^1	1×10^6
Hg-195	1×10^2	1×10^6
Hg-195m[a]	1×10^2	1×10^6
Hg-197	1×10^2	1×10^7
Hg-197m	1×10^2	1×10^6
Hg-199m	1×10^2	1×10^6
Hg-203	1×10^2	1×10^5
Tl-194	1×10^1	1×10^6
Tl-194m	1×10^1	1×10^6
Tl-195	1×10^1	1×10^6
Tl-197	1×10^2	1×10^6
Tl-198	1×10^1	1×10^6
Tl-198m	1×10^1	1×10^6
Tl-199	1×10^2	1×10^6
Tl-200	1×10^1	1×10^6
Tl-201	1×10^2	1×10^6
Tl-202	1×10^2	1×10^6
Tl-204	1×10^4	1×10^4
Pb-195m	1×10^1	1×10^6
Pb-198	1×10^2	1×10^6
Pb-199	1×10^1	1×10^6
Pb-200	1×10^2	1×10^6
Pb-201	1×10^1	1×10^6
Pb-202	1×10^3	1×10^6
Pb-202m	1×10^1	1×10^6
Pb-203	1×10^2	1×10^6
Pb-205	1×10^4	1×10^7
Pb-209	1×10^5	1×10^6
Pb-210[a]	1×10^1	1×10^4
Pb-211	1×10^2	1×10^6
Pb-212[a]	1×10^1	1×10^5
Pb-214	1×10^2	1×10^6
Bi-200	1×10^1	1×10^6
Bi-201	1×10^1	1×10^6
Bi-202	1×10^1	1×10^6
Bi-203	1×10^1	1×10^6
Bi-205	1×10^1	1×10^6
Bi-206	1×10^1	1×10^5
Bi-207	1×10^1	1×10^6
Bi-210	1×10^3	1×10^6
Bi-210m[a]	1×10^1	1×10^5

Radionuclide	Activity concentration (Bq/g)	Activity (Bq)
Bi-212[a]	1×10^{1}	1×10^{5}
Bi-213	1×10^{2}	1×10^{6}
Bi-214	1×10^{1}	1×10^{5}
Po-203	1×10^{1}	1×10^{6}
Po-205	1×10^{1}	1×10^{6}
Po-206	1×10^{1}	1×10^{6}
Po-207	1×10^{1}	1×10^{6}
Po-208	1×10^{1}	1×10^{4}
Po-209	1×10^{1}	1×10^{4}
Po-210	1×10^{1}	1×10^{4}
At-207	1×10^{1}	1×10^{6}
At-211	1×10^{3}	1×10^{7}
Fr-222	1×10^{3}	1×10^{5}
Fr-223	1×10^{2}	1×10^{6}
Rn-220[a]	1×10^{4}	1×10^{7}
Rn-222[a]	1×10^{1}	1×10^{8}
Ra-223[a]	1×10^{2}	1×10^{5}
Ra-224[a]	1×10^{1}	1×10^{5}
Ra-225	1×10^{2}	1×10^{5}
Ra-226[a]	1×10^{1}	1×10^{4}
Ra-227	1×10^{2}	1×10^{6}
Ra-228[a]	1×10^{1}	1×10^{5}
Ac-224	1×10^{2}	1×10^{6}
Ac-225[a]	1×10^{1}	1×10^{4}
Ac-226	1×10^{2}	1×10^{5}
Ac-227[a]	1×10^{-1}	1×10^{3}
Ac-228	1×10^{1}	1×10^{6}
Th-226[a]	1×10^{3}	1×10^{7}
Th-227	1×10^{1}	1×10^{4}
Th-228[a]	1×10^{0}	1×10^{4}
Th-229[a]	1×10^{0}	1×10^{3}
Th-230	1×10^{0}	1×10^{4}
Th-231	1×10^{3}	1×10^{7}
Th-232	1×10^{1}	1×10^{4}
Th-234[a]	1×10^{3}	1×10^{5}
Pa-227	1×10^{1}	1×10^{6}
Pa-228	1×10^{1}	1×10^{6}
Pa-230	1×10^{1}	1×10^{6}
Pa-231	1×10^{0}	1×10^{3}
Pa-232	1×10^{1}	1×10^{6}
Pa-233	1×10^{2}	1×10^{7}

Radionuclide	Activity concentration (Bq/g)	Activity (Bq)
Pa-234	1×10^{1}	1×10^{6}
U-230[a]	1×10^{1}	1×10^{5}
U-231	1×10^{2}	1×10^{7}
U-232[a]	1×10^{0}	1×10^{3}
U-233	1×10^{1}	1×10^{4}
U-234	1×10^{1}	1×10^{4}
U-235[a]	1×10^{1}	1×10^{4}
U-236	1×10^{1}	1×10^{4}
U-237	1×10^{2}	1×10^{6}
U-238[a]	1×10^{1}	1×10^{4}
U-239	1×10^{2}	1×10^{6}
U-240	1×10^{3}	1×10^{7}
U-240[a]	1×10^{1}	1×10^{6}
Np-232	1×10^{1}	1×10^{6}
Np-233	1×10^{2}	1×10^{7}
Np-234	1×10^{1}	1×10^{6}
Np-235	1×10^{3}	1×10^{7}
Np-236(1.5×10^{5} a)	1×10^{2}	1×10^{5}
Np-236(22.5 h)	1×10^{3}	1×10^{7}
Np-237[a]	1×10^{0}	1×10^{3}
Np-238	1×10^{2}	1×10^{6}
Np-239	1×10^{2}	1×10^{7}
Np-240	1×10^{1}	1×10^{6}
Pu-234	1×10^{2}	1×10^{7}
Pu-235	1×10^{2}	1×10^{7}
Pu-236	1×10^{1}	1×10^{4}
Pu-237	1×10^{3}	1×10^{7}
Pu-238	1×10^{0}	1×10^{4}
Pu-239	1×10^{0}	1×10^{4}
Pu-240	1×10^{0}	1×10^{3}
Pu-241	1×10^{2}	1×10^{5}
Pu-242	1×10^{0}	1×10^{4}
Pu-243	1×10^{3}	1×10^{7}
Pu-244	1×10^{0}	1×10^{4}
Pu-245	1×10^{2}	1×10^{6}
Pu-246	1×10^{2}	1×10^{6}
Am-237	1×10^{2}	1×10^{6}
Am-238	1×10^{1}	1×10^{6}
Am-239	1×10^{2}	1×10^{6}
Am-240	1×10^{1}	1×10^{6}
Am-241	1×10^{0}	1×10^{4}

Radionuclide	Activity concentration (Bq/g)	Activity (Bq)
Am-242	1×10^{3}	1×10^{6}
Am-242m[a]	1×10^{0}	1×10^{4}
Am-243[a]	1×10^{0}	1×10^{3}
Am-244	1×10^{1}	1×10^{6}
Am-244m	1×10^{4}	1×10^{7}
Am-245	1×10^{3}	1×10^{6}
Am-246	1×10^{1}	1×10^{5}
Am-246m	1×10^{1}	1×10^{6}
Cm-238	1×10^{2}	1×10^{7}
Cm-240	1×10^{2}	1×10^{5}
Cm-241	1×10^{2}	1×10^{6}
Cm-242	1×10^{2}	1×10^{5}
Cm-243	1×10^{0}	1×10^{4}
Cm-244	1×10^{1}	1×10^{4}
Cm-245	1×10^{0}	1×10^{3}
Cm-246	1×10^{0}	1×10^{3}
Cm-247	1×10^{0}	1×10^{4}
Cm-248	1×10^{0}	1×10^{3}
Cm-249	1×10^{3}	1×10^{6}
Cm-250	1×10^{-1}	1×10^{3}
Bk-245	1×10^{2}	1×10^{6}
Bk-246	1×10^{1}	1×10^{6}
Bk-247	1×10^{0}	1×10^{4}
Bk-249	1×10^{3}	1×10^{6}

Radionuclide	Activity concentration (Bq/g)	Activity (Bq)
Bk-250	1×10^{1}	1×10^{6}
Cf-244	1×10^{4}	1×10^{7}
Cf-246	1×10^{3}	1×10^{6}
Cf-248	1×10^{1}	1×10^{4}
Cf-249	1×10^{0}	1×10^{3}
Cf-250	1×10^{1}	1×10^{4}
Cf-251	1×10^{0}	1×10^{3}
Cf-252	1×10^{1}	1×10^{4}
Cf-253	1×10^{2}	1×10^{5}
Cf-254	1×10^{0}	1×10^{3}
Es-250	1×10^{2}	1×10^{6}
Es-251	1×10^{2}	1×10^{7}
Es-253	1×10^{2}	1×10^{5}
Es-254	1×10^{1}	1×10^{4}
Es-254m	1×10^{2}	1×10^{6}
Fm-252	1×10^{3}	1×10^{6}
Fm-253	1×10^{2}	1×10^{6}
Fm-254	1×10^{4}	1×10^{7}
Fm-255	1×10^{3}	1×10^{6}
Fm-257	1×10^{1}	1×10^{5}
Md-257	1×10^{2}	1×10^{7}
Md-258	1×10^{2}	1×10^{5}

[a] Parent radionuclides, and their progeny whose dose contributions are taken into account in the dose calculations (thus requiring only the exemption level of the parent radionuclide to be considered), are listed in the following.

Parent	Progeny
Ge-68	Ga-68
Rb-83	Kr-83m
Sr-82	Rb-82
Sr-90	Y-90
Y-87	Sr-87m
Zr-93	Nb-93m
Zr-97	Nb-97
Ru-106	Rh-106
Ag-108m	Ag-108
Sn-121m	Sn-121 (0.776)
Sn-126	Sb-126m
Xe-122	I-122
Cs-137	Ba-137m
Ba-140	La-140
Ce-134	La-134
Ce-144	Pr-144
Gd-146	Eu-146
Hf-172	Lu-172
W-178	Ta-178
W-188	Re-188
Re-189	Os-189m (0.241)
Ir-189	Os-189m
Pt-188	Ir-188
Hg-194	Au-194
Hg-195m	Hg-195 (0.542)
Pb-210	Bi-210, Po-210
Pb-212	Bi-212, Tl-208 (0.36), Po-212 (0.64)
Bi-210m	Tl-206
Bi-212	Tl-208 (0.36), Po-212 (0.64)
Rn-220	Po-216
Rn-222	Po-218, Pb-214, Bi-214, Po-214
Ra-223	Rn-219, Po-215, Pb-211, Bi-211, Tl-207
Ra-224	Rn-220, Po-216, Pb-212, Bi-212, Tl-208 (0.36), Po-212 (0.64)
Ra-226	Rn-222, Po-218, Pb-214, Bi-214, Po-214, Pb-210, Bi-210, Po-210
Ra-228	Ac-228
Ac-225	Fr-221, At-217, Bi-213, Po-213 (0.978), Tl-209 (0.0216), Pb-209 (0.978)
Ac-227	Fr-223 (0.0138)
Th-226	Ra-222, Rn-218, Po-214
Th-228	Ra-224, Rn-220, Po-216, Pb-212, Bi-212,Tl-208 (0.36), Po-212 (0.64)
Th-229	Ra-225, Ac-225, Fr-221, At-217, Bi-213, Po-213, Pb-209
Th-234	Pa-234m
U-230	Th-226, Ra-222, Rn-218, Po-214
U-232	Th-228, Ra-224, Rn-220, Po-216, Pb-212, Bi-212, Tl-208 (0.36), Po-212 (0.64)
U-235	Th-231
U-238	Th-234, Pa-234m
U-240	Np-240m
Np-237	Pa-233
Am-242m	Am-242
Am-243	Np-239

TABLE I-2. LEVELS FOR EXEMPTION OF BULK AMOUNTS OF SOLID MATERIAL WITHOUT FURTHER CONSIDERATION AND FOR CLEARANCE OF SOLID MATERIAL WITHOUT FURTHER CONSIDERATION: ACTIVITY CONCENTRATIONS OF RADIONUCLIDES OF ARTIFICIAL ORIGIN *(see footnote 58)*

Radionuclide	Activity concentration (Bq/g)	Radionuclide	Activity concentration (Bq/g)	Radionuclide	Activity concentration (Bq/g)
H-3	100	Co-60m	1000	Nb-95	1
Be-7	10	Co-61	100	Nb-97[a]	10
C-14	1	Co-62m	10	Nb-98	10
F-18	10	Ni-59	100	Mo-90	10
Na-22	0.1	Ni-63	100	Mo-93	10
Na-24	1	Ni-65	10	Mo-99[a]	10
Si-31	1000	Cu-64	100	Mo-101[a]	10
P-32	1000	Zn-65	0.1	Tc-96	1
P-33	1000	Zn-69	1000	Tc-96m	1000
S-35	100	Zn-69m[a]	10	Tc-97	10
Cl-36	1	Ga-72	10	Tc-97m	100
Cl-38	10	Ge-71	10 000	Tc-99	1
K-42	100	As-73	1000	Tc-99m	100
K-43	10	As-74	10	Ru-97	10
Ca-45	100	As-76	10	Ru-103[a]	1
Ca-47	10	As-77	1000	Ru-105[a]	10
Sc-46	0.1	Se-75	1	Ru-106[a]	0.1
Sc-47	100	Br-82	1	Rh-103m	10 000
Sc-48	1	Rb-86	100	Rh-105	100
V-48	1	Sr-85	1	Pd-103[a]	1000
Cr-51	100	Sr-85m	100	Pd-109[a]	100
Mn-51	10	Sr-87m	100	Ag-105	1
Mn-52	1	Sr-89	1000	Ag-110m[a]	0.1
Mn-52m	10	Sr-90[a]	1	Ag-111	100
Mn-53	100	Sr-91[a]	10	Cd-109[a]	1
Mn-54	0.1	Sr-92	10	Cd-115[a]	10
Mn-56	10	Y-90	1000	Cd-115m[a]	100
Fe-52[a]	10	Y-91	100	In-111	10
Fe-55	1000	Y-91m	100	In-113m	100
Fe-59	1	Y-92	100	In-114m[a]	10
Co-55	10	Y-93	100	In-115m	100
Co-56	0.1	Zr-93	10	Sn-113[a]	1
Co-57	1	Zr-95[a]	1	Sn-125	10
Co-58	1	Zr-97[a]	10	Sb-122	10
Co-58m	10 000	Nb-93m	10	Sb-124	1
Co-60	0.1	Nb-94	0.1	Sb-125[a]	0.1

Radionuclide	Activity concentration (Bq/g)	Radionuclide	Activity concentration (Bq/g)	Radionuclide	Activity concentration (Bq/g)
Te-123m	1	Nd-149	100	Hg-197	100
Te-125m	1000	Pm-147	1000	Hg-197m	100
Te-127	1000	Pm-149	1000	Hg-203	10
Te-127m[a]	10	Sm-151	1000	Tl-200	10
Te-129	100	Sm-153	100	Tl-201	100
Te-129m[a]	10	Eu-152	0.1	Tl-202	10
Te-131	100	Eu-152m	100	Tl-204	1
Te-131m[a]	10	Eu-154	0.1	Pb-203	10
Te-132[a]	1	Eu-155	1	Bi-206	1
Te-133	10	Gd-153	10	Bi-207	0.1
Te-133m	10	Gd-159	100	Po-203	10
Te-134	10	Tb-160	1	Po-205	10
I-123	100	Dy-165	1000	Po-207	10
I-125	100	Dy-166	100	At-211	1000
I-126	10	Ho-166	100	Ra-225	10
I-129	0.01	Er-169	1000	Ra-227	100
I-130	10	Er-171	100	Th-226	1000
I-131	10	Tm-170	100	Th-229	0.1
I-132	10	Tm-171	1000	Pa-230	10
I-133	10	Yb-175	100	Pa-233	10
I-134	10	Lu-177	100	U-230[b]	10
I-135	10	Hf-181	1	U-231[a]	100
Cs-129	10	Ta-182	0.1	U-232[a]	0.1
Cs-131	1000	W-181	10	U-233	1
Cs-132	10	W-185	1000	U-236	10
Cs-134	0.1	W-187	10	U-237	100
Cs-134m	1000	Re-186	1000	U-239	100
Cs-135	100	Re-188	100	U-240[a]	100
Cs-136	1	Os-185	1	Np-237[a]	1
Cs-137[a]	0.1	Os-191	100	Np-239	100
Cs-138	10	Os-191m	1000	Np-240	10
Ba-131	10	Os-193	100	Pu-234	100
Ba-140	1	Ir-190	1	Pu-235	100
La-140	1	Ir-192	1	Pu-236	1
Ce-139	1	Ir-194	100	Pu-237	100
Ce-141	100	Pt-191	10	Pu-238	0.1
Ce-143	10	Pt-193m	1000	Pu-239	0.1
Ce-144	10	Pt-197	1000	Pu-240	0.1
Pr-142	100	Pt-197m	100	Pu-241	10
Pr-143	1000	Au-198	10	Pu-242	0.1
Nd-147	100	Au-199	100	Pu-243	1000

Radionuclide	Activity concentration (Bq/g)
Pu-244[a]	0.1
Am-241	0.1
Am-242	1000
Am-242m[a]	0.1
Am-243[a]	0.1
Cm-242	10
Cm-243	1
Cm-244	1
Cm-245	0.1
Cm-246	0.1
Cm-247[a]	0.1
Cm-248	0.1
Bk-249	100
Cf-246	1000
Cf-248	1
Cf-249	0.1
Cf-250	1
Cf-251	0.1
Cf-252	1
Cf-253	100
Cf-254	1
Es-253	100
Es-254[a]	0.1
Es-254m[a]	10
Fm-254	10 000
Fm-255	100

[a] Parent radionuclides, and their progeny whose dose contributions are taken into account in the dose calculations (thus requiring only the exemption level of the parent radionuclide to be considered), are listed in the following.

Fe-52	Mn-52m
Zn-69m	Zn-69
Sr-90	Y-90
Sr-91	Y-91m
Zr-95	Nb-95
Zr-97	Nb-97m, Nb-97
Nb-97	Nb-97m
Mo-99	Tc-99m
Mo-101	Tc-101
Ru-103	Rh-103m
Ru-105	Rh-105m
Ru-106	Rh-106
Pd-103	Rh-103m
Pd-109	Ag-109m
Ag-110m	Ag-110
Cd-109	Ag-109m
Cd-115	In-115m
Cd-115m	In-115m
In-114m	In-114
Sn-113	In-113m
Sb-125	Te-125m
Te-127m	Te-127
Te-129m	Te-129
Te-131m	Te-131
Te132	I-132
Cs-137	Ba-137m
Ce-144	Pr-144, Pr-144m
U-232sec	Th-228, Ra-224, Rn-220, Po-216, Pb-212, Bi-212, Tl-208
U-240	Np-240m, Np-240
Np237	Pa-233
Pu-244	U-240, Np-240m, Np-240
Am-242m	Np-238
Am-243	Np-239
Cm-247	Pu-243
Es-254	Bk-250
Es-254m	Fm-254

TABLE I-3: LEVELS FOR CLEARANCE OF MATERIAL: ACTIVITY CONCENTRATIONS OF RADIONUCLIDES OF NATURAL ORIGIN

Radionuclide	Activity concentration (Bq/g)
K-40	10
Each radionuclide in the uranium and thorium decay chains	1

Schedule II
CATEGORIES FOR SEALED SOURCES USED IN COMMON PRACTICES

II-1. Table II-1 shows categories for sealed sources used in common practices, and Table II-2 shows activity corresponding to a dangerous source (D value) for selected radionuclides.

TABLE II-1. CATEGORIES FOR SEALED SOURCES USED IN COMMON PRACTICES

Category	Ratio of activity in the source to activity that is considered dangerous[i a] (A/D)	Example of sources[b] and practices
1	A/D ≥ 1000	Radioisotope thermoelectric generators (RTGs) Irradiators Teletherapy sources Fixed, multibeam teletherapy ('gamma knife') sources
2	1000>A/D ≥ 10	Industrial gamma radiography sources High/medium dose rate brachytherapy sources
3	10>A/D ≥ 1	Fixed industrial gauges incorporating high activity sources Well logging gauges
4	1>A/D ≥ 0.01	Low dose rate brachytherapy sources (except eye plaques and permanent implants) Industrial gauges not incorporating high activity sources Bone densitometers Static eliminators
5	0.01>A/D and A>Exempt[c]	Low dose rate brachytherapy eye plaques and permanent implant sources X ray fluorescence devices Electron capture devices Mossbauer spectrometry sources Positron emission tomography check sources

[i a] A is the activity of the radionuclide in a source and D is the activity of that radionuclide that is regarded as dangerous. A dangerous source is defined as one that could, if not under control, give rise to exposure sufficient to cause severe deterministic effects. Values of D for selected radionuclides are given in Table II-2 on the basis of the quantity of radioactive material that could give rise to severe deterministic effects for given exposure scenarios and for given dose criteria. This column of the table can thus be used to determine the category of a source, purely on the basis of the value of A/D. This may be appropriate if, for example: the practice is not known or is not listed; if sources have a short half-life and/or are unsealed; or if sources are aggregated.

[b] Factors other than A/D have been taken into consideration in assigning these sources to a particular category [29].

[c] Exempt quantities are given in Schedule I.

TABLE II-2. ACTIVITY[a] CORRESPONDING TO A DANGEROUS SOURCE (D VALUE[b]) FOR SELECTED RADIONUCLIDES

Radionuclide	D value (TBq)
Am-241	6×10^{-2}
Am-241/Be	6×10^{-2}
Au-198	2×10^{-1}
Cd-109	2×10^{1}
Cf-252	2×10^{-2}
Cm-244	5×10^{-2}
Co-57	7×10^{-1}
Co-60	3×10^{-2}
Cs-137	1×10^{-1}
Fe-55	8×10^{2}
Gd-153	1×10^{0}
Ge-68	7×10^{-2}
H-3	2×10^{3}
I-125	2×10^{-1}
I-131	2×10^{-1}
Ir-192	8×10^{-2}
Kr-85	3×10^{1}
Mo-99	3×10^{-1}

Radionuclide	D value (TBq)
Ni-63	6×10^{1}
P-32	1×10^{1}
Pd-103	9×10^{1}
Pm-147	4×10^{1}
Po-210	6×10^{-2}
Pu-238	6×10^{-2}
Pu-239/Be	6×10^{-2}
Ra-226	4×10^{-2}
Ru-106 (Rh-106)	3×10^{-1}
Se-75	2×10^{-1}
Sr-90 (Y-90)	1×10^{0}
Tc-99m	7×10^{-1}
Tl-204	2×10^{1}
Tm-170	2×10^{1}
Yb-169	3×10^{-1}

a Since Table II-2 does not show which dose criteria were used, these *D* values cannot be used 'in reverse' to derive possible doses due to sources of known activity.

b Full details of the derivation of the *D* values and *D* values for additional radionuclides are provided in Ref. [27].

Schedule III
DOSE LIMITS FOR PLANNED EXPOSURE SITUATIONS

OCCUPATIONAL EXPOSURE

III-1. For occupational exposure of workers over the age of 18 years, the dose limits are:

(a) An effective dose of 20 mSv per year averaged over five consecutive years[64] (100 mSv in 5 years), and of 50 mSv in any single year;
(b) An equivalent dose to the lens of the eye of 20 mSv per year averaged over 5 consecutive years (100 mSv in 5 years) and of 50 mSv in any single year;
(c) An equivalent dose to the extremities (hands and feet) or the skin[65] of 500 mSv in a year.

Additional restrictions apply to occupational exposure for a female worker who has notified pregnancy or is breast-feeding (para. 3.114).

III-2. For occupational exposure of apprentices of 16 to 18 years of age who are being trained for employment involving radiation and for exposure of students of age 16 to 18 who use sources in the course of their studies, the dose limits are:

(a) An effective dose of 6 mSv in a year;
(b) An equivalent dose to the lens of the eye of 20 mSv in a year;
(c) An equivalent dose to the extremities (hands and feet) or the skin[65] of 150 mSv in a year.

PUBLIC EXPOSURE

III-3. For public exposure, the dose limits are:

(a) An effective dose of 1 mSv in a year;
(b) In special circumstances[66], a higher value of effective dose in a single year could apply, provided that the average effective dose over five consecutive years does not exceed 1 mSv per year;
(c) An equivalent dose to the lens of the eye of 15 mSv in a year;
(d) An equivalent dose to the skin of 50 mSv in a year.

VERIFICATION OF COMPLIANCE WITH DOSE LIMITS

III-4. The effective dose limits specified here in Schedule III apply to the sum of the relevant doses from external exposure in the specified period and the relevant committed doses from intakes in the same period; the period for calculating the committed dose shall normally be 50 years for intakes by adults and up to age 70 years for intakes by children[67].

---

[64] The start of the averaging period shall be coincident with the first day of the relevant annual period after the date of entry into force of these Standards, with no retrospective averaging.

[65] The equivalent dose limits for the skin apply to the average dose over 1 cm^2 of the most highly irradiated area of the skin. The dose to the skin also contributes to the effective dose, this contribution being the average dose to the entire skin multiplied by the tissue weighting factor for the skin.

[66] For example, in authorized, justified and planned operational circumstances that lead to transitory increases in exposures.

[67] Information on procedures for the assessment of the effective dose to workers and members of the public is provided in IAEA Safety Guides and in ICRP publications.

III-5. For occupational exposure, the personal dose equivalent $H_p(10)$ may be used as an approximation of the effective dose from external exposure to penetrating radiation.

III-6. Values of the effective dose per unit air kerma free-in-air and per unit particle fluence are given in Tables III-1A–III-1D (see p.139), which is included in the PDF file in the accompanying CD-ROM attached to the inside back cover of the printed publication [29].

III-7. Doses per unit intake (dose coefficients) for the estimation of the committed effective dose for ingestion and inhalation of radionuclides are given in Tables III-2A–III-2H (see p.139), which is included in the PDF file in the accompanying CD-ROM attached to the inside back cover [30, 31].

Schedule IV
CRITERIA FOR USE IN EMERGENCY PREPAREDNESS AND RESPONSE

IV-1. Table IV-1 gives generic criteria for acute doses for which protective actions and other response actions are expected to be undertaken under any circumstances to avoid or to minimize severe deterministic effects.

IV-2. Table IV-2 gives guidance values for restricting exposure of emergency workers.

TABLE IV-1. GENERIC CRITERIA FOR ACUTE DOSES FOR WHICH PROTECTIVE ACTIONS AND OTHER RESPONSE ACTIONS ARE EXPECTED TO BE UNDERTAKEN UNDER ANY CIRCUMSTANCES TO AVOID OR TO MINIMIZE SEVERE DETERMINISTIC EFFECTS

External acute exposure (< 10 hours)		If the dose is projected:
$AD_{Red\ marrow}$ [a]	1 Gy	– Take precautionary urgent protective actions immediately (even under difficult conditions) to keep doses below the generic criteria
AD_{Fetus}	0.1 Gy	– Provide public information and warnings
AD_{Tissue} [b]	25 Gy at 0.5 cm	– Carry out urgent decontamination
AD_{Skin} [c]:	10 Gy to 100 cm^2	
Internal exposure from acute intake (Δ = 30 days[d])		If the dose has been received:
$AD(\Delta)_{Red\ marrow}$	0.2 Gy for radionuclides with atomic number $Z \geq 90$[e] 2 Gy for radionuclides with atomic number $Z \leq 89$[e]	– Perform immediate medical examination, consultation and indicated medical treatment – Carry out contamination control
$AD(\Delta)_{Thyroid}$	2 Gy	– Carry out immediate decorporation[f] (if applicable)
$AD(\Delta)_{Lung}$ [g]	30 Gy	– Carry out registration for long term health monitoring
$AD(\Delta)_{Colon}$	20 Gy	
$AD(\Delta')_{Fetus}$ [h]	0.1 Gy	– Provide comprehensive psychological counselling

[a] $AD_{Red\ marrow}$ represents the average RBE weighted absorbed dose to internal tissues or organs (e.g. red marrow, lung, small intestine, gonads, thyroid) and to the lens of the eye from exposure in a uniform field of strongly penetrating radiation.

[b] Dose delivered to 100 cm^2 at a depth of 0.5 cm under the body surface in tissue due to close contact with a radioactive source (e.g. source carried in the hand or pocket).

[c] The dose is to the 100 cm^2 dermis (skin structures at a depth of 40 mg/cm^2 (or 0.4 mm) below the surface).

[d] $AD(\Delta)$ is the RBE weighted absorbed dose delivered over the period of time Δ by the intake (I_{05}) that will result in a severe deterministic effect in 5% of exposed individuals.

[e] Different criteria are used to take account of the significant difference in the radionuclide specific intake threshold values for the radionuclides in these groups.

[f] The generic criterion for decorporation is based on the projected dose without decorporation. Decorporation is the biological processes, facilitated by a chemical or biological agent, by which incorporated radionuclides are removed from the human body.

[g] For the purposes of these generic criteria 'lung' means the alveolar-interstitial region of the respiratory tract.

[h] For this particular case, Δ' means the period of in utero development.

TABLE IV-2. GUIDANCE VALUES FOR RESTRICTING EXPOSURE OF EMERGENCY WORKERS

Tasks	Guidance value[a]
Life saving actions	$H_P(10)$[b] < 500 mSv This value may be exceeded under circumstances in which the expected benefits to others clearly outweigh the emergency worker's own health risks, and the emergency worker volunteers to take the action and understands and accepts this health risk
Actions to prevent severe deterministic effects and actions to prevent the development of catastrophic conditions that could significantly affect people and the environment	$H_P(10) < 500$ mSv
Actions to avert a large collective dose	$H_P(10) < 100$ mSv

[a] These values apply only for the dose from exposure to external penetrating radiation. Doses from exposure to non-penetrating external radiation and from intake or skin contamination need to be prevented by all possible means. If this is not feasible, the effective dose and the equivalent dose to an organ that are received have to be limited to minimize the health risk to the individual in line with the risk associated with the guidance values given here.

[b] $H_P(10)$ is the personal dose equivalent $H_P(d)$ where $d = 10$ mm.

REFERENCES

References are to editions that are current as of the time of publication of these Standards. Editions that supersede these may be adopted under national legislation. In the event that the publications referenced here are superseded, please refer to the most recent editions.
See also: http://www-ns.iaea.org/standards/

[1] INTERNATIONAL COMMISSION ON RADIOLOGICAL PROTECTION, The 2007 Recommendations of the International Commission on Radiological Protection, Publication 103, Pergamon Press, Oxford and New York (2007).

[2] EUROPEAN ATOMIC ENERGY COMMUNITY, FOOD AND AGRICULTURE ORGANIZATION OF THE UNITED NATIONS, INTERNATIONAL ATOMIC ENERGY AGENCY, INTERNATIONAL LABOUR ORGANIZATION, INTERNATIONAL MARITIME ORGANIZATION, OECD NUCLEAR ENERGY AGENCY, PAN AMERICAN HEALTH ORGANIZATION, UNITED NATIONS ENVIROMENT PROGRAMME, WORLD HEALTH ORGANIZATION, Fundamental Safety Principles, IAEA Safety Standards Series No. SF-1, IAEA, Vienna (2006).

[3] UNITED NATIONS, Sources and Effects of Ionizing Radiation (Report to the General Assembly), Scientific Committee on the Effects of Atomic Radiation (UNSCEAR), UN, New York (2000).

[4] UNITED NATIONS, Effects of Ionizing Radiation. Volume I: Report to the General Assembly, Scientific Annexes A and B; Volume II: Scientific Annexes C, D and E. United Nations Scientific Committee on the Effects of Atomic Radiation, UNSCEAR 2006 Report. United Nations sales publications E.08.IX.6 (2008) and E.09.IX.5 (2009), UN, New York..

[5] INTERNATIONAL COMMISSION ON RADIOLOGICAL PROTECTION, Statement on Radon, ICRP Ref 00/902/09, (2009).

[6] WORLD HEALTH ORGANIZATION, WHO Handbook on Indoor Radon: a Public Health Perspective, WHO, Geneva (2009).

[7] INTERNATIONAL COMMISSION ON RADIATION UNITS AND MEASUREMENTS, Quantities and Units in Radiation Protection Dosimetry (Report 51), ICRU, Bethesda (1993).

[8] INTERNATIONAL ATOMIC ENERGY AGENCY, Governmental, Legal and Regulatory Framework for Safety, IAEA Safety Standards Series No. GSR Part 1, IAEA, Vienna (2010).

[9] INTERNATIONAL ATOMIC ENERGY AGENCY, Decommissioning of Facilities Using Radioactive Material, IAEA Safety Standards Series No. WS-R-5, IAEA, Vienna (2006).

[10] INTERNATIONAL ATOMIC ENERGY AGENCY, Predisposal Management of Radioactive Waste, IAEA Safety Standards Series No. GSR Part 5, IAEA, Vienna (2009).

[11] INTERNATIONAL ATOMIC ENERGY AGENCY, Disposal of Radioactive Waste, IAEA Safety Standards Series No. SSR-5, IAEA, Vienna (2011)

[12] INTERNATIONAL ATOMIC ENERGY AGENCY, Regulations for the Safe Transport of Radioactive Material, 2009 Edition, IAEA Safety Standards Series No. TS-R-1, IAEA, Vienna (2009).

[13] INTERNATIONAL ATOMIC ENERGY AGENCY, The Management System for Facilities and Activities, IAEA Safety Standards Series No. GS-R-3, IAEA, Vienna (2006)

[14] INTERNATIONAL ATOMIC ENERGY AGENCY, Safety Assessment for Facilities and Activities, IAEA Safety Standards Series No. GSR Part 4, IAEA, Vienna (2009)

[15] FOOD AND AGRICULTURE ORGANIZATION OF THE UNITED NATIONS, INTERNATIONAL ATOMIC ENERGY AGENCY, INTERNATIONAL LABOUR ORGANIZATION, OECD NUCLEAR ENERGY AGENCY, PAN AMERICAN HEALTH ORGANIZATION, UNITED NATIONS OFFICE FOR THE CO-ORDINATION OF HUMANITARIAN AFFAIRS, WORLD HEALTH ORGANIZATION, Preparedness and Response for a Nuclear or Radiological Emergency, IAEA Safety Standards Series No. GS-R-2, IAEA, Vienna (2002).

[16] INTERNATIONAL ORGANIZATION FOR STANDARDIZATION, Basic Ionizing Radiation Symbol, ISO 361, ISO, Geneva (1975).

[17] INTERNATIONAL ORGANIZATION FOR STANDARDIZATION, Ionizing-Radiation Warning — Supplementary Symbol, ISO 21482, ISO, Geneva (2007).

[18] Council Directive 96/29 Euratom of 13 May 1996, Laying down basic safety standards for the protection of health of workers and the general public against the dangers arising from ionizing radiation, Official Journal of the European Communities No. L 159, Office for Official Publications of the European Communities, Luxembourg (1996).

[19] INTERNATIONAL LABOUR ORGANIZATION, Technical and Ethical Guidelines for Workers' Health Surveillance; Occupational Safety and Health Series 72, ILO, Geneva (1998).

[20] WORLD MEDICAL ASSOCIATION, 18th World Medical Assembly, Helsinki, 1974, as amended by the 59th World Medical Assembly, Seoul, 2008.

[21] COUNCIL FOR INTERNATIONAL ORGANIZATIONS OF MEDICAL SCIENCES in collaboration with WORLD HEALTH ORGANIZATION, International Ethical Guidelines for Biomedical Research Involving Human Subjects, CIOMS, Geneva (2002).

[22] INTERNATIONAL COMMISSION ON RADIOLOGICAL PROTECTION, Radiological Protection in Biomedical Research, ICRP Publication 62, Annals of the ICRP 22(3) (1991).

[23] JOINT FAO/WHO FOOD STANDARDS PROGRAMME, CODEX ALIMENTARIUS COMMISSION, Codex General Standard for Contaminants and Toxins in Foods, Schedule 1 — Radionuclides, CODEX STAN 193-1995, CAC, Rome (2006).

[24] WORLD HEALTH ORGANIZATION, Guidelines for Drinking-water Quality — 4th Ed., WHO, Geneva (2011).

[25] EUROPEAN COMMISSION, Principles and Methods for Establishing Concentrations and Quantities (Exemption Values) below which Reporting is not Required in the European Directive, Radiation Protection 65, Office for Official Publications of the European Communities, Luxembourg (1993) NATIONAL RADIOLOGICAL PROTECTION BOARD, Exempt Concentrations and Quantities for Radionuclides not included in the European Basic Safety Standards Directive, Mobbs, S.F., Harvey, M.P., NRPB –R306, Chilton, (1998).

[26] INTERNATIONAL ATOMIC ENERGY AGENCY, Derivation of Activity Concentration Values for Exclusion, Exemption and Clearance, Safety Reports Series No. 44, IAEA, Vienna (2005).

[27] INTERNATIONAL ATOMIC ENERGY AGENCY, Dangerous Quantities of Radioactive Material (D-values), EPR-D-VALUES 2006, IAEA, Vienna (2006).

[28] INTERNATIONAL ATOMIC ENERGY AGENCY, Categorization of Radioactive Sources, IAEA Safety Standards Series No. RS-G-1.9, IAEA, Vienna (2005).

[29] INTERNATIONAL COMMISSION ON RADIOLOGICAL PROTECTION, Conversion Coefficients for Use in Radiological Protection against External Radiation, ICRP Publication 74, Annals of the ICRP Vol. 26/3 (1997).

[30] INTERNATIONAL COMMISSION ON RADIOLOGICAL PROTECTION, ICRP Database of Dose Coefficients: Workers and Members of the Public, Version 2.0.1 (CD-ROM), Elsevier Science, Amsterdam (2003).

[31] INTERNATIONAL COMMISSION ON RADIOLOGICAL PROTECTION, Individual Monitoring for Internal Exposure of Workers, Publication 78, Annals of the ICRP Vol. 27/3-4, Replacement of ICRP Publication 54 (1998).

Annex

GENERIC CRITERIA FOR PROTECTIVE ACTIONS AND OTHER RESPONSE ACTIONS IN EMERGENCY EXPOSURE SITUATIONS TO REDUCE THE RISK OF STOCHASTIC EFFECTS

A-1. Table A-1 in this Annex provides a set of generic criteria (expressed in terms of projected dose and dose that has been received) for use in the protection strategy that are compatible with reference levels (expressed in terms of residual dose) within a range of 20–100 mSv, and provides details of specific protective actions and other response actions in different timeframes.

A-2. For the thyroid, iodine thyroid blocking is an urgent protective action that is prescribed: (i) if exposure due to radioactive iodine is involved, (ii) before or shortly after a release of radioactive iodine and (iii) only within a short period after the intake of radioactive iodine.

A-3. In the absence of national guidance, the generic criteria could be used as a basis for the development of criteria at the national level. In exceptional situations, the use of a higher value for the generic criteria may be necessary, such as when no replacement food or water is available.

TABLE A-1. GENERIC CRITERIA FOR PROTECTIVE ACTIONS AND OTHER RESPONSE ACTIONS IN EMERGENCY EXPOSURE SITUATIONS TO REDUCE THE RISK OF STOCHASTIC EFFECTS

Generic criteria		Examples of protective actions and other response actions
Projected dose that exceeds the following generic criteria: Take urgent protective actions and other response actions		
$H_{Thyroid}$	50 mSv in the first 7 days	Iodine thyroid blocking
E H_{Fetus}	100 mSv in the first 7 days 100 mSv in the first 7 days	Sheltering; evacuation; decontamination; restriction of consumption of food, milk and water; contamination control; public reassurance
Projected dose that exceeds the following generic criteria: Take protective actions and other response actions early in the response		
E H_{Fetus}	100 mSv per annum 100 mSv for the full period of in utero development	Temporary relocation; decontamination; replacement of food, milk and water; public reassurance
Dose that has been received and that exceeds the following generic criteria: Take longer term medical actions to detect and to effectively treat radiation induced health effects		
E	100 mSv in a month	Screening based on equivalent doses to specific radiosensitive organs (as a basis for medical follow-up), counselling
H_{Fetus}	100 mSv for the full period of in utero development	Counselling to allow informed decisions to be made in individual circumstances

DEFINITIONS

The following definitions apply for the purposes of these Standards.
Further definitions are provided in IAEA Safety Glossary: Terminology Used in Nuclear Safety and Radiation Protection (2007 Edition), IAEA, Vienna (2007).
http://www-ns.iaea.org/standards/safety-glossary
The symbol 'Θ' denotes an information note; this note is not part of the definition.

absorbed dose

The fundamental dosimetric quantity D, defined as:

$$D = \frac{\mathrm{d}\bar{\varepsilon}}{\mathrm{d}m}$$

where $\mathrm{d}\bar{\varepsilon}$ is the mean energy imparted by *ionizing radiation* to matter in a volume element and dm is the mass of matter in the volume element.

Θ The unit for *absorbed dose* is joule per kilogram (J/kg), given the name *gray (Gy).*

Θ The energy can be averaged over any defined volume, the average *dose* being equal to the total energy imparted in the volume divided by the mass in the volume.

Θ *Absorbed dose* is defined at a point; for the average *dose* in a tissue or organ, see *organ dose.*

accident

Any unintended *event*, including operating errors, equipment *failures* and other mishaps, the consequences or potential consequences of which are not negligible from the point of view of *protection* and *safety*.

activation

The *process* of inducing *radioactivity*.

activity

1. The quantity A for an amount of radionuclide in a given energy state at a given time, defined as:

$$A(t) = \frac{\mathrm{d}N}{\mathrm{d}t}$$

where dN is the expectation value of the number of spontaneous nuclear transformations from the given energy state in the time interval dt.

Θ The SI unit of activity is the reciprocal second (s^{-1}), termed the *becquerel* (Bq).

2. See *facilities and activities.*

ambient dose equivalent, H*(d)

The *dose equivalent* that would be produced by the corresponding aligned and expanded field in the *ICRU sphere* at a depth *d* on the radius opposing the direction of the aligned field.

Θ Parameter defined at a point in a *radiation* field. Used as a directly measurable proxy (i.e. substitute) for *effective dose* for use in *monitoring* of *external exposure*.

Θ The recommended value of *d* for *strongly penetrating radiation* is 10 mm.

annual dose

The *dose* from *external exposure* in a year plus the *committed dose* from *intakes* of radionuclides in that year.

approval

The granting of consent by a *regulatory body*.

area monitoring

A form of *workplace monitoring* in which an area is monitored by taking measurements at different points in that area.

Θ As opposed to measurements by a static monitor.

assessment

The *process*, and the result, of analysing systematically and evaluating the hazards associated with *sources* and *practices*, and associated *protection and safety* measures.

authorization

The granting by a *regulatory body* or other governmental body of written permission for a person or organization to conduct specified *activities*.

bioassay

Any *procedure* used to determine the nature, *activity*, location or retention of radionuclides in the body by direct (in vivo) measurement or by in vitro analysis of material excreted or otherwise removed from the body.

carers and comforters

Persons who willingly and voluntarily help (other than in their occupation) in the care, support and comfort of *patients* undergoing *radiological procedures* for medical diagnosis or medical treatment.

clearance

The removal of *regulatory control* by the *regulatory body* from *radioactive material* or *radioactive* objects within notified or authorized *practices*.

Θ Removal of *control* in this context refers to *control* applied for *radiation protection* purposes.

clearance level

A value, established by a *regulatory body* and expressed in terms of *activity concentration*, at or below which *regulatory control* may be removed from a *source* of *radiation* within a notified or authorized practice.

committed dose

The *lifetime dose* expected to result from an *intake*.

committed effective dose

The quantity $E(\tau)$, defined as:

$$E(\tau) = \sum_{T} w_T \cdot H_T(\tau)$$

where $H_T(\tau)$ is the *committed equivalent dose* to tissue T over the integration time τ and w_T is the *tissue weighting factor* for tissue T. When τ is not specified, it will be taken to be 50 years for adults and the time to age 70 years for *intakes* by children.

committed equivalent dose

The quantity $H_T(\tau)$, defined as:

$$H_T(\tau) = \int_{t_0}^{t_0+\tau} \dot{H}_T(t)\mathrm{d}t$$

where t_0 is the time of *intake*, $\dot{H}_T(t)$ is the *equivalent dose rate* at time t in organ or tissue T and τ is the time elapsed after an intake of *radioactive material*. When τ is not specified, it will be taken to be 50 years for adults and the time to age 70 years for *intakes* by children.

confinement

Prevention or *control* of releases of *radioactive material* to the environment in *operation* or in *accidents*.

constraint

A prospective and *source* related value of individual dose (*dose constraint*) or risk (*risk constraint*) that is used in planned exposure situations as a parameter for the optimization of protection and safety for the source, and that serves as a boundary in defining the range of options in optimization.

Θ For occupational exposure, a constraint on individual dose to workers established and used by registrants and licensees to set the range of options in optimizing protection and safety for the source.

Θ For public exposure, the dose constraint is a source related value established or approved by the government or the regulatory body, with account taken of the doses from planned operations of all sources under control. The dose constraint for each particular source is intended, among other things, to ensure that the sum of doses from planned operations for all sources under control remains within the dose limit.

Θ The risk constraint is a source related value that provides a basic level of protection for the individuals most at risk from a source. This risk is a function of the probability of an unintended event causing a dose, and the probability of the detriment due to the dose. Risk constraints correspond to dose constraints but apply to potential exposure.

Θ For medical exposure, the dose constraint is a source related value used in optimizing the protection of carers and comforters of patients undergoing radiological procedures, and the protection of volunteers subject to exposure as part of a programme of biomedical research.

consumer product

A device or manufactured item into which radionuclides have deliberately been incorporated or produced by activation, or which generates ionizing radiation, and which can be sold or made available to members of the public without special surveillance or regulatory control after sale.

Θ This includes items such as smoke detectors and luminous dials into which radionuclides have deliberately been incorporated and ion generating tubes. It does not include building materials, ceramic tiles, spa waters, minerals and foodstuffs and it excludes products and appliances installed in public places (e.g. exit signs).

containment

Methods or physical *structures* designed to prevent or *control* the release and the *dispersion* of *radioactive substances.*

contamination

Radioactive substances on surfaces, or within solids, liquids or gases (including the human body), where its presence is unintended or undesirable, or the process giving rise to its presence in such places.

Θ *Contamination* does not include residual *radioactive material* remaining at a site after the completion of *decommissioning.*

Θ The term *contamination* may have a connotation that is not intended. The term *contamination* refers only to the presence of *radioactivity*, and gives no indication of the magnitude of the hazard involved.

control

The function or power or (usually as *controls*) means of directing, regulating or restraining.

Θ It should be noted that the usual meaning of the English word *control* in *safety* related contexts is somewhat 'stronger' (more active) than that of its usual translations and other similar words in some other languages. For example, '*control*' typically implies not only checking or *monitoring* something but also ensuring that corrective or *enforcement* measures are taken if the results of the checking or *monitoring* indicate such a need. This is in contrast, for example, to the more limited usage of the equivalent word in French and Spanish.

regulatory control. Any form of *control* or regulation applied to *facilities and activities* by a *regulatory body* for reasons relating to *nuclear safety* and *radiation protection* or to *nuclear security.*

controlled area

A defined area in which specific *protection* measures and *safety* provisions are or could be required for controlling *exposures* or preventing the spread of *contamination* in normal working conditions, and preventing or limiting the extent of *potential exposures*.

decontamination

The complete or partial removal of *contamination* by a deliberate physical, chemical or biological process.

> Θ This definition is intended to include a wide range of processes for removing *contamination* from people, equipment and buildings, but to exclude the removal of radionuclides from within the human body or the removal of radionuclides by natural weathering or *migration processes*, which are not considered to be *decontamination*.

decorporation

The biological processes, facilitated by a chemical or biological agent, by which incorporated radionuclides are removed from the human body.

defence in depth

A hierarchical deployment of different levels of diverse equipment and *procedures* to prevent the escalation of *anticipated operational occurrences* and to maintain the effectiveness of physical *barriers* placed between a *source* or *radioactive material* and *workers*, *members of the public* or the environment, in *operational states* and, for some *barriers*, in *accident conditions*.

> Θ The objectives of defence in depth are:
> (a) To compensate for potential human and component failures;
> (b) To maintain the effectiveness of the barriers by averting damage to the facility and to the barriers themselves;
> (c) To protect *workers*, *members of the public* and the *environment* from harm in accident conditions in the event that these barriers are not fully effective.

deterministic effect

A *health effect* of *radiation* for which generally a threshold level of *dose* exists above which the severity of the effect is greater for a higher *dose*.

> Θ The level of the threshold *dose* is characteristic of the particular *health effect* but may also depend, to a limited extent, on the exposed individual. Examples of *deterministic effects* include erythema and acute radiation syndrome (radiation sickness).

> Θ Such an effect is described as a ***severe deterministic effect*** if it is fatal or life threatening or results in a permanent injury that reduces quality of life.

> Θ Deterministic effects are also referred to as 'harmful tissue reactions'.

diagnostic reference level

A level used in medical imaging to indicate whether, in routine conditions, the *dose* to the patient or the amount of *radiopharmaceuticals* administered in a specified *radiological procedure* is unusually high or unusually low for that procedure.

directional dose equivalent, H′(d,Ω)

The *dose equivalent* that would be produced by the corresponding expanded field in the *ICRU sphere* at a depth *d* on a radius in a specified direction *Ω*.

> Θ Parameter defined at a point in a *radiation* field. Used as a directly measurable proxy (i.e. substitute) for *equivalent dose* in the skin for use in *monitoring* of *external exposure*.
>
> Θ The recommended value of d for weakly penetrating radiation is 0.07 mm.

disposal

Emplacement of *waste* in an appropriate *facility* without the intention of retrieval.

dose

1. A measure of the energy deposited by *radiation* in a target.

2. *Absorbed dose*, *committed equivalent dose*, *committed effective dose*, *effective dose*, *equivalent dose* or *organ dose*, as indicated by the context.

committed dose. *committed equivalent dose* or *committed effective dose*.

dose assessment

Assessment of the *dose(s)* to an individual or group of people.

dose constraint

See *constraint*.

dose limit

The value of the *effective dose* or the *equivalent dose* to individuals in *planned exposure situations* that is not to be exceeded.

effective dose

The quantity *E*, defined as a summation of all the tissue *equivalent doses*, each multiplied by the appropriate *tissue weighting factor*:

$$E = \sum_{\mathrm{T}} w_{\mathrm{T}} \cdot H_{\mathrm{T}}$$

where H_{T} is the *equivalent dose* in tissue T and w_{T} is the *tissue weighting factor* for tissue T. From the definition of *equivalent dose*, it follows that:

$$E = \sum_{\mathrm{T}} w_{\mathrm{T}} \cdot \sum_{\mathrm{R}} w_{\mathrm{R}} \cdot D_{\mathrm{T,R}}$$

where w_R is the *radiation weighting factor* for *radiation* R and $D_{T,R}$ is the average *absorbed dose* in the tissue or organ T.

Θ The unit of *effective dose* is joule per kilogram (J/kg), given the name *sievert* (Sv).An explanation of the quantity is given in Annex B of ICRP 103 [1].

Θ *Effective dose* is a measure of *dose* designed to reflect the amount of *radiation detriment* likely to result from the *dose*.

Θ *Effective dose* cannot be used to quantify higher *doses* or to make decisions on the need for any medical treatment relating to *deterministic effects*.

Θ Values of *effective dose* from any type(s) of *radiation* and mode(s) of *exposure* can be compared directly.

emergency

A non-routine situation that necessitates prompt action, primarily to mitigate a hazard or adverse consequences for human health and *safety*, quality of life, property or the *environment*. This includes *nuclear or radiation emergencies* and conventional *emergencies* such as fires, release of hazardous chemicals, storms or earthquakes. It includes situations for which prompt action is warranted to mitigate the effects of a perceived hazard.

nuclear or radiation emergency. An *emergency* in which there is, or is perceived to be, a hazard due to:

(a) The energy resulting from a nuclear chain reaction or from the decay of the products of a chain reaction; or

(b) *Radiation exposure*.

emergency action level, EAL

A specific, predetermined, observable criterion used to detect, recognize and determine the *emergency class*.

emergency class

A set of conditions that warrant a similar immediate emergency response.

Θ This is the term used for communicating to the *response organizations* and to members of the public the level of response needed. The *events* that belong to a given *emergency class* are defined by criteria specific to the installation, *source* or *practice*, which, if exceeded, indicate classification at the prescribed level. For each *emergency class*, the initial actions of the *response organizations* are predefined.

emergency exposure situation

An emergency exposure situation is a situation of exposure that arises as a result of an accident, a malicious act, or any other unexpected event, and requires prompt action in order to avoid or reduce adverse consequences.

Θ Emergency exposures can be reduced only by protective actions and other response actions.

emergency plan

A description of the objectives, policy and concept of *operations* for the response to an *emergency* and of the *structure*, authorities and responsibilities for a systematic, coordinated and effective response. The *emergency plan* serves as the basis for the development of other plans, *procedures* and checklists.

emergency preparedness

The capability to take actions that will effectively mitigate the consequences of an *emergency* for human health and *safety*, quality of life, property and the environment.

emergency procedures

A set of instructions describing in detail the actions to be taken by response personnel in an *emergency*.

emergency response

The performance of actions to mitigate the consequences of an *emergency* for human health and *safety*, quality of life, property and the environment. It may also provide a basis for the resumption of normal social and economic activity.

emergency response arrangements

The integrated set of infrastructural elements necessary to provide the capability for performing a specified function or task *required* in response to a *nuclear or radiation emergency*. These elements may include authorities and responsibilities, organization, coordination, personnel, plans, *procedures*, *facilities*, equipment or training.

emergency worker

A person having specified duties as a worker in response to an *emergency*.

Θ Emergency workers may include workers employed by registrants and licensees as well as personnel of responding organizations, such as police officers, firefighters, medical personnel, and drivers and crews of evacuation vehicles.

employer

A person or organization with recognized responsibilities, commitments and duties towards a *worker* in the employment of the person or organization by virtue of a mutually agreed relationship. (A self-employed person is regarded as being both an *employer* and a *worker*.)

environment

The conditions under which people, animals and plants live or develop and which sustain all life and development; especially such conditions as affected by human activities.

Θ ***Protection of the environment*** includes the protection and conservation of: non-human species, both animal and plant, and their biodiversity; environmental goods and services such as the production of food and feed; resources used in agriculture, forestry, fisheries and tourism; amenities used in spiritual, cultural and recreational activities; media such as soil, water and air; and natural processes such as carbon, nitrogen and water cycles.

environmental monitoring

The measurement of *external dose* rates due to *sources* in the environment or of radionuclide concentrations in environmental media.

Θ Contrasted with *source monitoring*.

equilibrium equivalent concentration

The *activity concentration* of ^{222}Rn or ^{220}Rn in *radioactive equilibrium* with its short lived progeny that would have the same *potential alpha energy* concentration as the actual (non-equilibrium) mixture.

Θ The *equilibrium equivalent concentration* of ^{222}Rn is given by

$$EEC\ ^{222}\mathrm{Rn} = 0.104 \times C(^{218}\mathrm{Po}) + 0.514 \times C(^{214}\mathrm{Pb}) + 0.382 \times C(^{214}\mathrm{Bi})$$

where $C(x)$ is the activity concentration of nuclide x in air. 1 Bq/m^3 *EEC* ^{222}Rn corresponds to 5.56×10^{-6} mJ/m^3.

Θ The *equilibrium equivalent concentration* of ^{220}Rn is given by

$$EEC\ ^{220}\mathrm{Rn} = 0.913 \times C(^{212}\mathrm{Pb}) + 0.087 \times C(^{212}\mathrm{Bi})$$

where $C(x)$ is the activity concentration of nuclide x in air. 1 Bq/m^3 *EEC* ^{220}Rn corresponds to 7.57×10^{-5} mJ/m^3.

equilibrium factor

The ratio of the *equilibrium equivalent activity concentration* of ^{222}Rn to the actual ^{222}Rn activity concentration.

equivalent dose

***equivalent dose, H_T*.** The quantity $H_{\mathrm{T,R}}$, defined as:

$$H_{\mathrm{T,R}} = w_{\mathrm{R}} \cdot D_{\mathrm{T,R}}$$

where $D_{\mathrm{T,R}}$ is the *absorbed dose* delivered by *radiation* type R averaged over a tissue or organ T and w_{R} is the *radiation weighting factor* for *radiation* type R. When the *radiation* field is composed of different *radiation* types with different values of w_{R} the *equivalent dose* is:

$$H_{\mathrm{T}} = \sum_{\mathrm{R}} w_{\mathrm{R}} \cdot D_{\mathrm{T,R}}$$

Θ The unit of equivalent dose is the sievert (Sv), equal to 1 J/kg. An explanation of the quantity is given in Annex B of ICRP 103 [1].

Θ *Equivalent dose* is a measure of the *dose* to a tissue or organ designed to reflect the amount of harm caused.

Θ *Equivalent dose* cannot be used to quantify higher doses or to make decisions on the need for any medical treatment relating to deterministic effects.

Θ Values of *equivalent dose* to a specified tissue from any type(s) of *radiation* can be compared directly.

evacuation

The rapid, temporary removal of people from an area to avoid or reduce short term *radiation exposure* in an *emergency*.

Θ Evacuation is an *urgent protective action*. If people are removed from the area for a longer period of time (more than a few months) the term *relocation* is used.

event

In the context of the reporting and *analysis* of *events*, an *event* is any occurrence unintended by the *operator*, including operating error, equipment *failure* or other mishap, and deliberate action on the part of others, the consequences or potential consequences of which are not negligible from the point of view of *protection* or *safety*.

exemption

The determination by a *regulatory body* that a *source* or *practice* need not be subject to some or all aspects of *regulatory control* on the basis that the *exposure* and the *potential exposure* due to the *source* or *practice* are too small to warrant the application of those aspects or that this is the optimum option for *protection* irrespective of the actual level of the *doses* or *risks*.

exemption level.

A value, established by a *regulatory body* and expressed in terms of *activity concentration*, total *activity*, *dose rate* or *radiation* energy, at or below which a *source* of *radiation* need not be subject to some or all aspects of *regulatory control*.

existing exposure situation

An existing exposure situation is a situation of exposure that already exists when a decision on the need for control needs to be taken.

Θ Existing exposure situations include exposure to natural background radiation that is amenable to control; exposure due to residual radioactive material that arose from past practices that were never subject to regulatory control or exposure due to residual radioactive material arising from a nuclear or radiation emergency after an emergency exposure situation has been declared to be ended.

exposure

The state or condition of being subject to irradiation.

external exposure. *Exposure* to *radiation* from a *source* outside the body.

internal exposure. *Exposure* to *radiation* from a *source* within the body.

exposure pathway

A route by which *radiation* or radionuclides can reach humans and cause *exposure*.

facilities and activities[68]

A general term encompassing *nuclear facilities*, uses of all *sources* of *ionizing radiation*, all *radioactive waste management activities*, *transport* of *radioactive material* and any other *practice* or circumstances in which people may be subject to exposure to *radiation* from naturally occurring or artificial *sources*.

Θ ***Facilities*** includes: *nuclear facilities*; *irradiation installations*; some mining and raw material processing *facilities* such as uranium mines; *radioactive waste management facilities*; and any other places where *radioactive material* is produced, processed, used, handled, stored or disposed of — or where *radiation generators* are installed — on such a scale that consideration of *protection and safety* is *required*.

Activities includes: the production, use, import and export of *radiation sources* for industrial, research and medical purposes; the *transport* of *radioactive material*; the *decommissioning* of *facilities*; *radioactive waste management activities* such as the *discharge* of effluents; and some aspects of the *remediation* of sites affected by residues from past *activities*.

Θ This term is intended to provide an alternative to the terminology of *sources* and *practices* (or *intervention*) to refer to general categories of situations. For example, a *practice* may involve many different *facilities and/or activities*, whereas the general definition (1) of *source* is too broad in some cases: a *facility or activity* might constitute a *source*, or might involve the use of many *sources*, depending upon the interpretation used.

Θ The term *facilities and activities* is very general, and includes those for which little or no *regulatory control* may be necessary or achievable: the more specific terms ***authorized facility*** and ***authorized activity*** should be used to distinguish those *facilities and activities* for which any form of *authorization* has been given.

Θ In the Fundamental Safety Principles (Safety Fundamentals), the term '*facilities and activities — existing and new — utilized for peaceful purposes*' is abbreviated for convenience to *facilities and activities* as a general term encompassing any human activity that may cause people to be exposed to *radiation risks* arising from naturally occurring or artificial *sources* (see Ref. [2], para. 1.9).

feed

Any single or multiple materials, whether processed, semi-processed or raw, that is intended to be fed directly to *food* producing animals.

fluence

Θ A measure of the strength of a *radiation* field. Commonly used without qualification to mean *particle fluence*.

energy fluence, Ψ. A measure of the energy density of a *radiation* field, defined as:

---

[68] A small number of 'catch-all' terms — namely: *facilities and activities*; *protection and safety*; and *structures, systems and components* — are defined in the IAEA Safety Glossary. These terms may be used in exactly the form listed to describe a whole group of things without cumbersome repetition, or slight variations of the terms may be used to refer to particular subgroups. Although the definitions include an indication of the meanings of the separate elements of the terms, these are not intended to be applied rigidly: if precise reference is needed to particular items covered by the catch-all term, more precise terms should be used.

$$\Psi = \frac{\mathrm{d}R}{\mathrm{d}a}$$

where dR is the *radiation* energy incident on a sphere of cross-sectional area da.

Θ The *energy fluence* rate

$$\frac{\mathrm{d}\Psi}{\mathrm{d}t}$$

is denoted by a lower case ψ.

Θ See INTERNATIONAL COMMISSION ON RADIOLOGICAL PROTECTION, Conversion Coefficients for Use in Radiological Protection against External Radiation, ICRP Publication 74, Annals of the ICRP Volume 26/3, Pergamon Press, Oxford and New York (1997).

particle fluence, Φ. A measure of the density of particles in a *radiation* field, defined as:

$$\Phi = \frac{\mathrm{d}N}{\mathrm{d}a}$$

where dN is the number of particles incident on a sphere of cross-sectional area da.

Θ The particle fluence rate

$$\frac{\mathrm{d}\Phi}{\mathrm{d}t}$$

is denoted by a lower case ϕ.

Θ See INTERNATIONAL COMMISSION ON RADIOLOGICAL PROTECTION, Conversion Coefficients for Use in Radiological Protection against External Radiation, ICRP Publication 74, Annals of the ICRP Volume 26/3, Pergamon Press, Oxford and New York (1997).

food

Any substance, whether processed, semi-processed or raw, which is intended for human consumption.

Θ This includes drink (other than freshwater), chewing gum and substances used in the preparation or processing of food; it does not include cosmetics, tobacco or drugs. Consumption in this context refers to ingestion.

graded approach

For a system of *control*, such as a regulatory system or a *safety system*, a *process* or method in which the stringency of the *control* measures and conditions to be applied is commensurate, to the extent practicable, with the likelihood and possible consequences of, and the level of *risk* associated with, a loss of *control*.

hazard assessment

Assessment of hazards associated with *facilities*, *activities* or *sources* within or beyond the borders of a State in order to identify:

(a) Those *events* and the associated areas for which *protective actions* may be *required* within the State;

(b) The actions that would be effective in mitigating the consequences of such *events*.

health authority

A governmental entity (at the national, regional or local level) that is responsible for policies and interventions, including the development of standards and the provision of guidance, for maintaining or improving human health, and that has the legal power of enforcing such policies and interventions.

health professional

An individual who has been formally recognized through appropriate national *procedures* to practise a profession related to health (e.g. medicine, dentistry, chiropractic, podiatry, nursing, medical physics, medical radiation technology, radiopharmacy, occupational health).

health screening programme

A programme in which a health test or medical examination is performed for the purpose of the early detection of disease.

health surveillance

See *workers' health surveillance*.

incident

Any unintended *event*, including operating errors, equipment *failures*, *initiating events*, *accident precursors*, *near misses* or other mishaps, or unauthorized act, *malicious* or non-malicious, the consequences or potential consequences of which are not negligible from the point of view of *protection and safety*.

individual monitoring

Monitoring using measurements by equipment worn by individual *workers*, or measurements of quantities of *radioactive material* in or on the bodies of individual workers, or measurement of radioactive material excreted by individual workers.

Θ Usually contrasted with *workplace monitoring*.

inspection imaging device

An imaging device designed specifically for imaging persons or cargo conveyances for the purpose of detecting concealed objects on or within the human body or within cargo or a vehicle.

Θ In some types of inspection imaging device ionizing radiation is used to produce images by backscatter, transmission or both. Other types of inspection imaging device utilize imaging by means of electrical and magnetic fields, ultrasound and sonar waves, nuclear magnetic resonance, microwaves, terahertz rays, millimetre waves, infrared radiation or visible light.

intake

1. The act or *process* of taking radionuclides into the body by inhalation or ingestion or through the skin.

2. The *activity* of a radionuclide taken into the body in a given time period or as a result of a given *event*.

interested party

A person, company, etc., with a concern or interest in the activities and performance of an organization, business, system, etc.

Θ The term *interested party* is used in a broad sense to mean a person or group having an interest in the performance of an organization. Those who can influence events may effectively become interested parties — whether their 'interest' is regarded as 'genuine' or not — in the sense that their views need to be considered. Interested parties have typically included the following: customers, owners, *operators*, employees, *suppliers*, partners, trade unions; the regulated industry or professionals; scientific bodies; governmental agencies or regulatory bodies (national, regional and local) whose responsibilities may cover nuclear energy; the media; members of the public (individuals, community groups and interest groups); and other States, especially neighbouring States that have entered into agreements providing for an exchange of information concerning possible transboundary impacts, or States involved in the export or import of certain technologies or materials.

investigation level

The value of a quantity such as *effective dose*, *intake* or *contamination* per unit area or volume at or above which an investigation would be conducted.

ionizing radiation

See *radiation*.

justification

1. The *process* of determining for a *planned exposure situation* whether a *practice* is, overall, beneficial; i.e., whether the expected benefits to individuals and to society from introducing or continuing the *practice* outweigh the harm (including *radiation detriment*) resulting from the *practice*.

2. The *process* of determining for an *emergency exposure situation* or an *existing exposure situation* whether a proposed *protective action* or *remedial action* is likely, overall, to be beneficial; i.e., whether the expected benefits to individuals and to society (including the reduction in *radiation detriment*) from introducing or continuing the *protective action* or *remedial action* outweigh the cost of such action and any harm or damage caused by the action.

kerma, K

The quantity K, defined as:

$$K = \frac{\mathrm{d}E_{\mathrm{tr}}}{\mathrm{d}m}$$

where $\mathrm{d}E_{\mathrm{tr}}$ is the sum of the initial kinetic energies of all charged ionizing particles liberated by uncharged ionizing particles in a material of mass $\mathrm{d}m$.

Θ SI unit is joules per kilogram (J/kg), termed *gray* (Gy).

Θ Originally an acronym for kinetic energy released in matter, but now accepted as a word.

air kerma. The *kerma* value for air.

Θ Under charged particle equilibrium conditions, the *air kerma* (in *gray*) is numerically approximately equal to the *absorbed dose* in air (in *gray*).

reference air kerma rate. The *kerma* rate to air, in air, at a reference distance of 1 m, corrected for air *attenuation* and scattering.

Θ This quantity is expressed in μGy/h at 1 m.

licence

A legal document issued by the *regulatory body* granting *authorization* to perform specified *activities* relating to a *facility or activity*.

Θ A *licence* is a product of the *authorization* process and a *practice* with a current *licence* is an authorized *practice*.

Θ *Authorization* may take other forms, such as *registration*.

Θ The *licensee* is the person or organization having overall responsibility for a facility or activity.

licensee

The holder of a current *licence*.

limit

The value of a quantity used in certain specified *activities* or circumstances that must not be exceeded.

authorized limit. A *limit* on a measurable quantity, established or formally accepted by a *regulatory body*.

operational limits and conditions. A set of rules setting forth parameter *limits*, the functional capability and the performance levels of equipment and personnel approved by the *regulatory body* for safe *operation* of an *authorized facility*.

linear energy transfer (LET), L_Δ

Defined generally as:

$$L_\Delta = \left(\frac{\mathrm{d}E}{\mathrm{d}\ell}\right)_\Delta$$

where dE is the energy lost in traversing distance dℓ and Δ is an upper bound on the energy transferred in any single collision.

Θ A measure of how, as a function of distance, energy is transferred from *radiation* to the exposed matter. A high value of *linear energy transfer* indicates that energy is deposited within a small distance.

Θ L_∞ (i.e. with $\Delta = \infty$) is termed the unrestricted linear energy transfer in defining *quality factor*.

Θ L_Δ is also known as the restricted linear collision stopping power.

management system

A set of interrelated or interacting elements (the system) for establishing policies and objectives and enabling the objectives to be achieved in an efficient and effective manner.

Θ The component parts of the management system include the organizational structure, resources and organizational processes. Management is defined (in ISO 9000) as coordinated activities to direct and control an organization.

Θ The management system integrates all elements of an organization into one coherent system to enable all of the organization's objectives to be achieved. These elements include the organizational structure, resources and processes. Personnel, equipment and organizational culture as well as the documented policies and processes are parts of the management system. The organization's processes have to address the totality of the requirements on the organization as established in, for example, IAEA safety standards and other international codes and standards.

medical exposure

Exposure incurred by patients for the purposes of medical or dental diagnosis or treatment; by *carers and comforters*; and by volunteers subject to *exposure* as part of a programme of biomedical research.

Θ A ***patient*** is an individual who is a recipient of services of health care professionals and/or their agents that are directed at (1) health promotion; (2) prevention of illness and injury; (3) monitoring health; (4) maintaining health; and (5) medical treatment of diseases, disorders and injuries in order to achieve a cure or, failing that, optimum comfort and function. Some asymptomatic individuals are included. For the purpose of these Standards, the term 'patient' refers only to those individuals undergoing radiological procedures.

medical physicist

A *health professional,* with specialist education and training in the concepts and techniques of applying physics in medicine, and competent to practise independently in one or more of the subfields (specialties) of medical physics.

Θ Competence of persons is normally assessed by the State by having a formal mechanism for registration, accreditation or certification of medical physicists in the various specialties (e.g. diagnostic radiology, radiation therapy, nuclear medicine). States that have yet to develop such a mechanism would need to assess the education, training and competence of any individual proposed by the licensee to act as a medical physicist and to decide, on the basis either of international accreditation standards or standards of a State where such an accreditation system exists, whether such an individual could undertake the functions of a medical physicist, within the required specialty.

medical radiation facility

A medical *facility* in which *radiological procedures* are carried out.

medical radiation technologist

A *health professional*, with specialist education and training in medical radiation technology, competent to carry out *radiological procedures,* on delegation from the *radiological medical practitioner,* in one or more of the specialties of medical radiation technology.

Θ Competence of persons is normally assessed by the State by having a formal mechanism for registration, accreditation or certification of medical radiation technologists in the various specialties (e.g. diagnostic radiology, radiation therapy, nuclear medicine). States that have yet to develop such a mechanism would need to assess the education, training and competence of any individual proposed by the licensee to act as a medical radiation technologist and to decide, on the basis either of international standards or standards of a State where such a system exists, whether such an individual could undertake the functions of a medical radiation technologist, within the required specialty.

medical radiological equipment

Radiological equipment used in *medical radiation facilities* to perform radiological procedures that either delivers an exposure to a person or directly controls or influences the extent of such exposure. The term applies to radiation generators, such as X ray machines or medical linear accelerators; to devices containing sealed sources, such as cobalt-60 teletherapy units; and to devices used in medical imaging to capture images, such as a gamma camera, image intensifier, flat panel detector or positron emission tomography scanner.

member of the public

For *protection and safety* purposes, in a general sense, any individual in the population except when subject to *occupational exposure* or *medical exposure*. For the purpose of verifying compliance with the *annual dose limit* for *public exposure*, this is the *representative person*.

monitoring

The measurement of *dose, dose rate* or *activity* related to the *assessment* or *control* of *exposure* to *radiation* or *radioactive substances*, and the interpretation of the results.

Θ 'Measurement' is used somewhat loosely here. The 'measurement' of *dose* often means the measurement of a *dose equivalent quantity* as a proxy (i.e. substitute) for a *dose quantity* that cannot be measured directly. Also, sampling may be involved as a preliminary step to measurement.

Θ *Monitoring* may be subdivided in two different ways: according to where the measurements are made, into *individual monitoring*, *workplace monitoring*, *source monitoring* and *environmental monitoring*; and, according to the purpose of the *monitoring*, into *routine monitoring*, *task related monitoring* and *special monitoring*.

natural background

The *doses*, *dose rates* or *activity concentrations* associated with *natural sources,* or any other *sources* in the environment that are not amenable to *control*.

Θ This is normally considered to include doses, dose rates or activity concentrations associated with natural sources, global fallout (but not local fallout) from atmospheric nuclear weapon tests and the Chernobyl accident.

natural source

A naturally occurring *source* of *radiation*, such as the sun and stars (*sources* of cosmic *radiation*) and rocks and soil (terrestrial *sources* of *radiation*), or any other material whose *radioactivity* is for all intents and purposes due only to radionuclides of natural origin, such as products or residues from the processing of minerals; but excluding radioactive material for use in a nuclear installation and radioactive waste generated in such an installation.

notification

A document submitted to the *regulatory body* by a person or organization to notify an intention to carry out a *practice* or other use of a *source*.

nuclear fuel cycle

All *operations* associated with the production of nuclear energy.

Θ This includes:

(a) Mining and processing of uranium or thorium ores;
(b) Enrichment of uranium;
(c) Manufacture of *nuclear fuel*;
(d) Operation of nuclear reactors (including *research reactors*);
(e) *Reprocessing* of *spent fuel*;
(f) All *waste management activities* (including *decommissioning*) relating to *operations* associated with the production of nuclear energy;
(g) Any related research and development *activities*.

nuclear installation

A *nuclear fuel* fabrication plant, *research reactor* (including subcritical and *critical assemblies*), nuclear power plant, *spent fuel storage facility*, enrichment plant or *reprocessing facility*.

Θ This is essentially any *authorized facilities* that are part of the *nuclear fuel cycle* except *facilities* for the mining or processing of uranium or thorium ores and *radioactive waste management facilities*.

nuclear or radiation emergency

See *emergency*.

(nuclear) security

The prevention and detection of, and response to, theft, *sabotage*, unauthorized access, illegal transfer or other *malicious* acts involving *nuclear material*, other *radioactive material* or their associated *facilities*.

Θ There is not an exact distinction between the general terms *safety* and *security*. In general, *security* is concerned with *malicious* or negligent actions by humans that could cause or threaten harm to other humans; *safety* is concerned with the broader issue of harm to humans (or the environment) from *radiation*, whatever the cause. The precise interaction between *security* and *safety* depends on the context. *Security* of *nuclear material* for reasons relating to non-proliferation is outside the scope of the IAEA safety standards.

occupancy factor

A typical fraction of the time for which a location is occupied by an individual or group.

occupational exposure

Exposure of *workers* incurred in the course of their work.

operational intervention level (OIL)

A set *level* of a measurable quantity that corresponds to a generic criterion.

Θ *Operational intervention levels* are typically expressed in terms of *dose rates* or of *activity* of *radioactive material* released, time integrated air *activity* concentrations, ground or surface concentrations, or *activity* concentrations of radionuclides in environmental, food or water samples. An *operational intervention level* is used immediately and directly (without further *assessment*) to determine the appropriate *protective actions* on the basis of an environmental measurement.

optimization of protection and safety

The process of determining what level of protection and safety would result in the magnitude of individual doses, the number of individuals (workers and members of the public) subject to exposure and the likelihood of exposure being "as low as reasonably achievable, economic and social factors being taken into account" (ALARA).

For medical exposures of patients, the optimization of protection and safety is the management of the radiation dose to the patient commensurate with the medical purpose.

Θ "protection and safety is optimized" means that optimization of protection and safety has been applied and the result of that process has been implemented.

personal dose equivalent, $H_p(d)$.

The *dose equivalent* in soft tissue below a specified point on the body at an appropriate depth *d*.

Θ Parameter used as a directly measurable proxy (i.e. substitute) for *equivalent dose* in tissues or organs or (with $d = 10$ mm) for *effective dose*, in *individual monitoring* of *external exposure*.

Θ The recommended values of *d* are 10 mm for *strongly penetrating radiation* and 0.07 mm for *weakly penetrating radiation*.

Θ 'Soft tissue' is commonly interpreted as the *ICRU sphere*.

planned exposure situation

A planned exposure situation is a situation of exposure that arises from the planned operation of a source or from a planned activity that results in an exposure from a source.

Θ Since provision for protection and safety can be made before embarking on the activity concerned, associated exposures and their probabilities of occurrence can be restricted from the outset. The primary means of controlling exposure in planned exposure situations is by good design of installations, equipment and operating procedures. In planned exposure situations, a certain level of exposure is expected to occur.

planning target volume

A geometrical concept used in radiation therapy for planning medical treatment with consideration of the net effect of movements of the patient and of the tissues to be irradiated, variations in size and shape of the tissues, and variations in beam geometry such as beam size and beam direction.

potential exposure

Prospective *exposure* that is not expected to be delivered with certainty but that may result from an *anticipated operational occurrence*, *accident* at a *source* or owing to an *event* or sequence of *events* of a probabilistic nature, including equipment *failures* and operating errors.

> Θ Potential exposure includes prospectively considered exposures from a source due to an event or sequence of events of a probabilistic nature, including those resulting from an accident, equipment failures, operating errors, natural phenomena (such as hurricanes, earthquakes and floods) and inadvertent human intrusion (such as the intrusion into a near surface waste disposal facility after institutional control is removed).

practice

Any human activity that introduces additional *sources* of *exposure* or additional *exposure pathways,* or modifies the network of *exposure pathways* from existing *sources*, so as to increase the *exposure* or the likelihood of *exposure* of people or the number of people exposed.

> ! *Radioactive waste* is generated as a result of *practices* that involve some beneficial effect, such as the generation of electricity by nuclear means or the diagnostic application of radioisotopes. The management of this *waste* is therefore only one part of the overall *practice*.

projected dose

The *dose* that would be expected to be received if planned protective actions were not taken.

protection (against radiation)

radiation protection (also ***radiological protection***). The *protection* of people from harmful effects of *exposure* to *ionizing radiation*, and the means for achieving this.

protection and safety

The *protection* of people against *exposure* to *ionizing radiation* or due to *radioactive material* and the *safety* of *sources*, including the means for achieving this, and the means for preventing *accidents* and for mitigating the consequences of *accidents* if they do occur.

> Θ For the purposes of the IAEA safety standards, '*protection and safety*' includes the *protection* of people against ionizing *radiation* and *radiation safety*; it does not include non-radiation-related aspects of *safety*. *Protection and safety* is concerned with both *radiation risks* under normal circumstances and *radiation risks* as a consequence of *incidents*, as well as with other possible direct consequences of a loss of *control* over a nuclear reactor core, nuclear chain reaction, *radioactive source* or any other *source* of radiation. *Safety measures* include actions to prevent *incidents* and arrangements put in place to mitigate their consequences if they were to occur.

protection of the environment

See *environment.*

protective action

An action for the purposes of avoiding or reducing *doses* that might otherwise be received in an *emergency exposure situation* or an *existing exposure situation.*

longer term protective action. A *protective action* that is not an *urgent protective action.*

Θ Such protective actions are likely to be prolonged over weeks, months or years.

Θ These include measures such as relocation, agricultural countermeasures and remedial actions.

mitigatory action. Immediate action by the *operator* or other party:

(1) To reduce the potential for conditions to develop that would result in *exposure* or a release of *radioactive material* requiring *emergency actions* on or off the site; or

(2) To mitigate *source* conditions that may result in *exposure* or a release of *radioactive material* requiring *emergency actions* on or off the site.

precautionary urgent protective action. *A protective action* in the event of a *nuclear or radiation emergency* which must be taken before or shortly after a release of *radioactive material,* or before an *exposure,* on the basis of the prevailing conditions to prevent or to reduce the *risk* of *severe deterministic effects.*

urgent protective action. A *protective action* in the event of an *emergency* which must be taken promptly (usually within hours) in order to be effective, and the effectiveness of which will be markedly reduced if it is delayed.

public exposure

Exposure incurred by *members of the public* due to *sources* in *planned exposure situations, emergency exposure situations* and *existing exposure situations*, excluding any *occupational exposure* or *medical exposure*.

qualified expert

An individual who, by virtue of certification by appropriate boards or societies, professional licences or academic qualifications and experience, is duly recognized as having expertise in a relevant field of specialization, e.g. medical physics, *radiation protection*, occupational health, fire safety, *quality management* or any relevant engineering or *safety* specialty.

quality assurance (QA)

The function of a *management system* that provides confidence that specified *requirements* will be fulfilled.

Θ Planned and systematic actions are necessary to provide adequate confidence that an item, *process* or service will satisfy given *requirements* for quality; for example, those specified in the *licence*. This statement is slightly modified from that in ISO 921:1997 (Nuclear Energy: Vocabulary) to say 'an item, *process* or service' instead of 'a product or service' and to add the example. A more general definition of *quality assurance* and definitions of related terms can be found in ISO 8402:1994.

radiation

! When used in IAEA publications, the term *radiation* normally refers only to *ionizing radiation*. The IAEA has no statutory responsibilities in relation to non-ionizing radiation.

ionizing radiation. For the purposes of *radiation protection*, *radiation* capable of producing ion pairs in biological material(s).

Θ *Ionizing radiation* can be divided into *low linear energy transfer radiation* and *high linear energy transfer radiation* (as a guide to its *relative biological effectiveness*), or into *strongly penetrating radiation* and *weakly penetrating radiation* (as an indication of its ability to penetrate shielding or the human body).

radiation detriment

The total harm that would eventually be incurred by a group that is subject to *exposure* and by its descendants as a result of the group's *exposure* to *radiation* from a *source*.

radiation generator

A device capable of generating *ionizing radiation*, such as X rays, neutrons, electrons or other charged particles, that may be used for scientific, industrial or medical purposes.

radiation protection

See *protection*.

radiation protection officer

A person technically competent in *radiation protection* matters relevant for a given type of *practice* who is designated by the *registrant, licensee* or employer to oversee the application of relevant *requirements*.

radiation risks

- Detrimental *health effects* of *exposure* to *radiation* (including the likelihood of such effects occurring).
- Any other *safety* related *risks* (including those to the environment) that might arise as a direct consequence of:
  - *Exposure* to *radiation*;
  - The presence of *radioactive material* (including *radioactive waste*) or its release to the environment;
  - A loss of *control* over a nuclear reactor core, nuclear chain reaction, *radioactive source* or any other *source* of *radiation*.

radiation weighting factor, w_R

A number by which the *absorbed dose* in a tissue or organ is multiplied to reflect the *relative biological effectiveness* of the *radiation* in inducing *stochastic effects* at low *doses*, the result being the *equivalent dose*.

RECOMMENDED RADIATION WEIGHTING FACTORS

Radiation type	w_R
Photons	1
Electrons and muons	1
Protons and charged pions	2
Alpha particles, fission fragments, heavy ions	20
Neutrons	A continuous function of neutron energy: $w_R = \begin{cases} 2.5 + 18.2\, e^{-[\ln(E_n)]^2/6}, & E_n < 1\ MeV \\ 5.0 + 17.0\, e^{-[\ln(2E_n)]^2/6}, & 1\ MeV \leq E_n \leq 50\ MeV \\ 2.5 + 3.25\, e^{-[\ln(0.04E_n)]^2/6}, & E_n > 50\ MeV \end{cases}$

Note: All values relate to the radiation incident on the body or, for internal radiation sources, emitted from the incorporated radionuclide(s).

radioactive (*adjective*)

1. Exhibiting *radioactivity*; emitting or relating to the emission of *ionizing radiation* or particles.

Θ This is the 'scientific' definition, and should not be confused with the 'regulatory' definition (2).

2. Designated in national law or by a *regulatory body* as being subject to *regulatory control* because of its *radioactivity*.

Θ This is the 'regulatory' definition, and should not be confused with the 'scientific' definition (1).

radioactive material

Material designated in national law or by a *regulatory body* as being subject to *regulatory control* because of its *radioactivity*.

Θ This is the 'regulatory' meaning of *radioactive* (2), and should not be confused with the 'scientific' meaning (1) of *radioactive*: 'exhibiting *radioactivity*; emitting or relating to the emission of *ionizing radiation* or particles'. The 'scientific' meaning of *radioactive* — as in ***radioactive substance*** — refers only to the presence of radioactivity, and gives no indication of the magnitude of the hazard involved.

radioactive source

A *source* containing *radioactive material* that is used as a source of radiation.

radioactive substance

Θ This is the 'scientific' meaning of *radioactive* (1), and should not be confused with the 'regulatory' meaning (2) of *radioactive*: 'Designated in national law or by a *regulatory body* as being subject to *regulatory control* because of its *radioactivity*.' The 'scientific' meaning of *radioactive* refers only to the presence of radioactivity, and gives no indication of the magnitude of the hazard involved.

radioactive waste

For legal and regulatory purposes, material for which no further use is foreseen that contains, or is contaminated with, radionuclides at activity concentrations or *activities* greater than *clearance levels* as established by the *regulatory body*.

! It should be recognized that this definition is purely for regulatory purposes, and that material with *activity concentrations* equal to or less than *clearance levels* is *radioactive* from a physical viewpoint, although the associated radiological hazards are considered negligible.

radioactive waste management

All administrative and operational *activities* involved in the handling, *pretreatment*, *treatment*, *conditioning*, *transport*, *storage* and *disposal* of *radioactive waste*.

predisposal. Any *waste management* steps carried out prior to *disposal*, such as *pretreatment*, *treatment*, *conditioning*, *storage* and *transport activities*.

Θ *Predisposal* is used as a contraction of 'pre-disposal *radioactive waste management*', not a form of *disposal*.

processing. Any *operation* that changes the characteristics of *waste*, including *pretreatment*, *treatment* and *conditioning*.

radioactive waste management facility

Facility specifically designed to handle, treat, condition, temporarily store or permanently dispose of *radioactive waste*.

radiological medical practitioner

A *health professional* with specialist education and training in the medical uses of radiation, who is competent to perform independently or to oversee procedures involving *medical exposure* in a given specialty.

Θ Competence of persons is normally assessed by the State by having a formal mechanism for registration, accreditation or certification of radiological medical practitioners in the given specialty (e.g. radiology, radiation therapy, nuclear medicine, dentistry, cardiology, etc.). States that have yet to develop such a mechanism need to assess the education, training and competence of any individual proposed by the licensee to act as a radiological medical practitioner and to decide, on the basis either of international standards or standards of a State where such a system exists, whether such an individual can undertake the functions of a radiological medical practitioner, within the required specialty.

radiological procedure

A medical imaging procedure or therapeutic procedure that involves *ionizing radiation*, such as a procedure in diagnostic radiology, nuclear medicine or radiation therapy, or any planning procedure, image guided interventional procedure or other interventional procedure involving radiation, delivered by a *radiation generator*, by a device containing a *sealed source* or by an *unsealed source*, or delivered by means of a radiopharmaceutical administered to a patient.

radiopharmacist

A *health professional*, with specialist education and training in radiopharmacy, who is competent to prepare and dispense radiopharmaceuticals used for the purposes of medical diagnosis and therapy.

Θ Competence of persons is normally assessed by the State by having a formal mechanism for registration, accreditation or certification of radiopharmacists. States that have yet to develop such a mechanism need to assess the education, training and competence of any individual proposed by the licensee to act as a radiopharmacist and to decide, on the basis either of international standards or standards of a State where such a system exists, whether such an individual can undertake the functions of a radiopharmacist.

radon

Any combination of isotopes of the element radon.

Θ For the purposes of these Standards, radon refers to radon-220 and radon-222.

radon progeny

The short lived *radioactive* decay products of radon-220 and radon-222.

Θ For radon-222, this includes the decay chain up to but not including lead-210, namely polonium-218, lead-214, bismuth-214 and polonium-214, plus traces of astatine-218, thallium-210 and lead-209. Lead-210, which has a *half-life* of 22.3 years, and its *radioactive* progeny — bismuth-210 and polonium-210, plus traces of mercury-206 and thallium-206 — are, strictly, progeny of radon-222, but they are not included in this listing because they will not normally be present in significant amounts in airborne form. For radon-220, this includes polonium-216, lead-212, bismuth-212, polonium-212 and thallium-208.

RBE weighted absorbed dose, AD_T

The quantity $AD_{T,R}$, defined as:

$$AD_{T,R} = D_{T,R} \times RBE_{T,R}$$

where $D_{T,R}$ is the *absorbed dose* delivered by *radiation of* type R averaged over a tissue or organ T and $RBE_{T,R}$ is the *relative biological effectiveness* for *radiation of* type R in the production of severe deterministic effects in a tissue or organ T. When the *radiation* field is composed of different *radiation* types with different values of $RBE_{T,R}$, the *RBE weighted absorbed dose* is given by:

$$AD_T = \sum_R D_{T,R} \times RBE_{T,R}$$

Θ The unit of *RBE weighted absorbed dose* is the *gray (Gy),* equal to 1 J/kg.

Θ *RBE weighted absorbed dose* is a measure of the *dose* to a tissue or organ designed to reflect the risk of development of severe deterministic effects.

Θ Values of *RBE weighted absorbed dose* to a specified tissue from any type(s) of *radiation* can be compared directly.

recording level

A level of *dose*, *exposure* or *intake* specified by the *regulatory body* at or above which values of *dose* to, *exposure* of or *intake* by *workers* are to be entered in their individual *exposure* records.

reference level

In an *emergency exposure situation* or an *existing exposure situation*, the level of *dose*, *risk* or *activity concentration* above which it is not appropriate to plan to allow *exposures* to occur and below which optimization of protection and safety would continue to be implemented.

Θ The chosen value for a *reference level* will depend upon the prevailing circumstances for the exposure under consideration.

referring medical practitioner

A *health professional* who, in accordance with national requirements, may refer individuals to a *radiological medical practitioner* for *medical exposure*.

registration

A form of *authorization* for *practices* of low or moderate *risks* whereby the person or organization responsible for the *practice* has, as appropriate, prepared and submitted a *safety assessment* of the *facilities* and equipment to the *regulatory body*. The *practice* or use is authorized with conditions or limitations as appropriate.

Θ The requirements for *safety assessment* and the conditions or limitations applied to the *practice* would be less severe for *registration* than those for licensing.

Θ Typical *practices* that are amenable to *registration* are those for which: (a) *safety* can largely be ensured by the *design* of the *facilities* and equipment; (b) the operating *procedures* are simple to follow; (c) the *safety* training requirements are minimal; and (d) there is a history of few problems with *safety* in *operations*. *Registration* is best suited to those *practices* for which *operations* do not vary significantly.

registrant.

The holder of a current *registration*.

Θ Other derivative terms should not be needed; a *registration* is a product of the *authorization process*, and a *practice* with a current *registration* is an authorized *practice*.

regulatory body

An authority or a system of authorities designated by the government of a State as having legal authority for conducting the regulatory *process*, including issuing *authorizations*, and thereby regulating *nuclear*, *radiation*, *radioactive waste* and *transport safety*.

Θ The national *competent authority* for the regulation of *radioactive material transport safety* (see Ref. [12]) is included in this description.

regulatory control

Θ See *control* (1).

relative biological effectiveness (RBE)

A measure of the relative effectiveness of different *radiation* types at inducing a specified *health effect,* expressed as the inverse ratio of the *absorbed doses* of two different *radiation* types that would produce the same degree of a defined biological *end point.*

Θ Values of relative biological effectiveness in causing the development of deterministic effects are selected to be representative of the severe deterministic effects that are significant to emergency preparedness and response. The tissue specific and radiation specific values of $RBE_{T,R}$ for the development of selected severe deterministic effects are as shown in the table.

Health effect	Critical organ	Exposure[a]	$RBE_{T,R}$
Haematopoietic syndrome	Red marrow	External and internal γ	1
		External and internal n	3
		Internal β	1
		Internal α	2
Pneumonitis	Lung[b]	External and internal γ	1
		External and internal n	3
		Internal β	1
		Internal α	7
Gastrointestinal syndrome	Colon	External and internal γ	1
		External and internal n	3
		Internal β	1
		Internal α	0[c]
Necrosis	Tissue[d]	External β, γ	1
		External n	3
Moist desquamation	Skin[e]	External β, γ	1
		External n	3
Hypothyroidism	Thyroid	Intake of iodine isotopes[f]	0.2
		Other thyroid seekers	1

[a]External β, γ exposure includes exposure due to bremsstrahlung produced within the material of the source.

[b]Tissue of the alveolar-interstitial region of the respiratory tract.

[c]For alpha emitters uniformly distributed in the contents of the colon, it is assumed that irradiation of the walls of the intestine is negligible.

[d]Tissue at a depth of 5 mm below the skin surface over an area of more than 100 cm^2.

[e]Tissue at a depth of 0.4 mm below the skin surface over an area of more than 100 cm^2.

[f]Uniform irradiation of the tissue of the thyroid gland is considered to be five times more likely to produce deterministic effects than internal exposure due to low energy beta emitting isotopes of iodine such as ^{131}I, ^{129}I, ^{125}I, ^{124}I and ^{123}I. Thyroid seeking radionuclides have a heterogeneous distribution in thyroid tissue. The isotope ^{131}I emits low energy beta particles, which leads to a reduced effectiveness of irradiation of critical thyroid tissue owing to the dissipation of the energy of the particles within other tissues.

remedial action

The removal of a *source* or the reduction of its magnitude (in terms of activity or amount) for the purposes of preventing or reducing *exposures* that might otherwise occur in an *existing exposure situation*.

remediation

Any measures that may be carried out to reduce the *radiation exposure* due to existing *contamination* of land areas through actions applied to the *contamination* itself (the *source*) or to the *exposure pathways* to humans.

Θ Complete removal of the *contamination* is not implied.

Θ See *decontamination*.

representative person

An individual receiving a *dose* that is representative of the *doses* to the more highly *exposed* individuals in the population.

Θ ICRP Publication 101 indicates that the dose to the representative person "is the equivalent of, and replaces, the mean dose in the 'critical group'", and provides guidance on assessing doses to the *representative person*. The concept of critical group remains valid.

Θ See *member of the public*.

residual dose

The *dose* expected to be incurred in the future after *protective actions* have been terminated (or a decision has been taken not to implement *protective actions*).

Θ This applies in an existing *exposure situation* or an *emergency exposure situation*.

response organization

An organization designated or otherwise recognized by a State as being responsible for managing or implementing any aspect of an *emergency response*.

risk

A multiattribute quantity expressing hazard, danger or chance of harmful or injurious consequences associated with exposures or *potential exposures*. It relates to quantities such as the probability that specific deleterious consequences may arise and the magnitude and character of such consequences.

risk constraint

See *constraint*.

safety

See *protection and safety*.

safety assessment

Assessment of all aspects of a *practice* that are relevant to *protection and safety*; for an *authorized facility*, this includes *siting*, *design* and *operation* of the *facility*.

safety culture

The assembly of characteristics and attitudes in organizations and individuals which establishes that, as an overriding priority, *protection and safety* issues receive the attention warranted by their significance.

safety measure

Any action that might be taken, condition that might be applied or *procedure* that might be followed to fulfil the basic *requirements* of Safety Requirements.

safety standards

Standards of *safety* issued pursuant to Article III(A)(6)[69] of the Statute of the IAEA.

Θ Requirements, regulations, standards, rules, codes of practice or recommendations established to protect people and the environment against ionizing radiation and to minimize danger to life and property.

scenario

A postulated or assumed set of conditions and/or *events*.

Θ Most commonly used in analysis or assessment to represent possible future conditions and/or events to be modelled, such as possible accidents at a nuclear facility, or the possible future evolution of a repository and its surroundings. A scenario may represent the conditions at a single point in time or a single event, or a time history of conditions and/or events (including processes).

Θ See *event*.

sealed source

A *radioactive source* in which the *radioactive material* is (a) permanently sealed in a capsule or (b) closely bonded and in a solid form.

security

See *(nuclear) security*.

source

1. Anything that may cause *radiation exposure* — such as by emitting *ionizing radiation* or by releasing *radioactive material* — and can be treated as a single entity for *protection and safety* purposes.

---

[69] "[The Agency is authorized…] To establish or adopt, in consultation and, where appropriate, in collaboration with the competent organs of the United Nations and with the specialized agencies concerned, standards of safety for protection of health and minimization of danger to life and property (including such standards for labour conditions)…"

Θ For example, materials emitting *radon* are *sources* in the environment; a sterilization gamma irradiation unit is a *source* for the *practice* of *radiation* preservation of food and sterilization of other products; an X ray unit may be a *source* for the *practice* of radiodiagnosis; a nuclear power plant is part of the *practice* of generating electricity by nuclear fission, and may be regarded as a *source* (e.g. with respect to *discharges* to the environment) or as a collection of *sources* (e.g. for occupational *radiation protection* purposes). A complex or multiple installation situated at one location or site may, as appropriate, be considered a single *source* for the purposes of application of international *safety standards*.

natural source. A naturally occurring *source* of *radiation*, such as the sun and stars (*sources* of cosmic *radiation*) and rocks and soil (terrestrial *sources* of *radiation*), or any other material whose *radioactivity* is for all intents and purposes due only to radionuclides of natural origin, such as products or residues from the processing of minerals; but excluding radioactive material for use in a nuclear installation and radioactive waste generated in such an installation.

radiation generator. A device capable of generating *ionizing radiation*, such as X rays, neutrons, electrons or other charged particles, that may be used for scientific, industrial or medical purposes.

2. *Radioactive material* used as a *source* of *radiation*.

Θ Such as those sources used for medical applications or in industrial instruments. These are, of course, *sources* as defined in (1), but this usage is less general.

dangerous source. A *source* that could, if not under *control*, give rise to *exposure* sufficient to cause *severe deterministic effects*. This categorization is used for determining the need for *emergency response arrangements* and is not to be confused with categorizations of *sources* for other purposes.

radioactive source. A *source* containing *radioactive material* that is used as a source of radiation.

sealed source. A *radioactive source* in which the *radioactive material* is (a) permanently sealed in a capsule or (b) closely bonded and in a solid form.

unsealed source. A *radioactive source* in which the *radioactive material* is neither (a) permanently sealed in a capsule nor (b) closely bonded and in a solid form.

source monitoring

The *measurement* of *activity* in *radioactive material* being released to the environment or of *external dose* rates due to *sources* within a *facility or activity*.

Θ Contrasted with *environmental monitoring*.

spent fuel

Nuclear fuel removed from a reactor following irradiation that is no longer usable in its present form because of depletion of *fissile material*, *poison* buildup or *radiation* damage.

Θ The adjective 'spent' suggests that *spent fuel* cannot be used as *fuel* in its present form (e.g. as in *spent source*). In practice, however, *spent fuel* is commonly used to refer to *fuel* which has been used as *fuel* but will no longer be used, whether or not it could be used (and which might more accurately be termed 'disused *fuel*').

standards dosimetry laboratory

A laboratory, designated by the relevant national authority, that possesses certification or accreditation necessary for the purpose of developing, maintaining or improving primary or secondary standards for radiation dosimetry.

stochastic effect

A *radiation* induced *health effect*, the probability of occurrence of which is greater for a higher *radiation dose* and the severity of which (if it occurs) is independent of *dose*.

Θ *Stochastic effects* may be somatic effects or hereditary effects, and generally occur without a threshold level of dose. Examples include solid cancers and leukaemia.

storage

The holding of *radioactive sources*, *spent fuel* or *radioactive waste* in a *facility* that provides for their/its *containment*, with the intention of retrieval.

structures, systems and components

A general term encompassing all of the elements (items) of a *facility* or *activity* which contribute to *protection and safety*, except *human factors*.

Θ ***Structures*** are the passive elements: buildings, vessels, shielding, etc. A ***system*** comprises several *components*, assembled in such a way as to perform a specific (active) function. A ***component*** is a discrete element of a *system*. Examples of components are wires, transistors, integrated circuits, motors, relays, solenoids, pipes, fittings, pumps, tanks and valves.

supervised area

A defined area not designated as a *controlled area* but for which *occupational exposure* conditions are kept under review, even though no specific *protection* measures or *safety* provisions are not normally needed.

supplier (of a source)

Any person or organization to whom a registrant or licensee delegates duties, totally or partially, in relation to the design, manufacture, production or construction of a source.

Θ The term 'supplier' includes designers, manufacturers, producers, constructors, assemblers, installers, distributors, sellers, exporters or importers of a source.

survey

radiological survey. An evaluation of the radiological conditions and potential hazards associated with the production, use, transfer, release, *disposal* or presence of *radioactive material* or other *sources* of *radiation*.

system

See *structures, systems and components*.

tissue weighting factor, w_T

Multiplier of the *equivalent dose* to an tissue or organ, as given by the *System of Radiological Protection*, used for *radiation protection* purposes to account for the different sensitivities of different tissues or organs to the induction of *stochastic effects* of *radiation*.

Θ Recommended tissue weighting factors:

Tissue	w_T	$\sum w_T$
Bone-marrow (red), Colon, Lung, Stomach, Breast, Remainder tissues*	0.12	0.72
Gonads	0.08	0.08
Bladder, Oesophagus, Liver, Thyroid	0.04	0.16
Bone surface, Brain, Salivary glands, Skin	0.01	0.04
	Total	1.00

* The w_T for the remainder tissues (0.12) applies to the arithmetic mean dose of the 13 organs and tissues for each sex listed below. Remainder tissues: adrenals, extrathoracic (ET) region, gall bladder, heart, kidneys, lymphatic nodes, muscle, oral mucosa, pancreas, prostate (male), small intestine, spleen, thymus, uterus/cervix (female).

Θ Recommended tissue weighting factors are included on the accompanying CD.

transboundary exposure

Exposure of *members of the public* in one State due to *radioactive material* released via *accidents*, *discharges* or *waste disposal* in another State.

transport

1. The deliberate physical movement of *radioactive material* (other than that forming part of the means of propulsion) from one place to another.

2. The movement of something as a result of being carried by a medium.

Θ A general term used when a number of different *processes* are involved. The most common examples are heat *transport* — a combination of *advection*, convection, etc., in a cooling medium — and radionuclide *transport* in the environment — which could include *processes* such as *advection*, *diffusion*, *sorption* and *uptake*.

trigger

A level or condition that is selected to act as an initiator for setting off an event or action (especially a response).

unsealed source

A *radioactive source* in which the *radioactive material* is neither (a) permanently sealed in a capsule nor (b) closely bonded and in a solid form.

urgent protective action

See *protective action*.

worker

Any person who works, whether full time, part time or temporarily, for an *employer* and who has recognized rights and duties in relation to occupational *radiation protection*.

Θ A self-employed person is regarded as having the duties of both an *employer* and a *worker*.

workers' health surveillance

Medical supervision intended to ensure the initial and continuing fitness of *workers* for their intended tasks.

workplace monitoring

Monitoring using measurements made in the working environment.

Θ Usually contrasted with *individual monitoring*.

CONTRIBUTORS TO DRAFTING AND REVIEW

Abu-Eid, R.	Nuclear Regulatory Commission, United States of America
Ahier, B.	OECD Nuclear Energy Agency (NEA)
Akhadi, M.	National Nuclear Energy Agency (BATAN), Indonesia
Al-Arfaj, A.	Institute of Atomic Energy Research (AERI), Saudi Arabia
Ali, H.	Atomic Licensing Board (AELB), Malaysia
Ali, M.	Pakistan Nuclear Regulatory Authority (PNRA), Pakistan
Amaral, E.	International Atomic Energy Agency
Amor Calvo, I.	Consejo de Seguridad Nuclear (CSN), Spain
Ampuero Flores, C.	Instituto Peruano de Energía Nuclear (IPEN), Peru
Andersen, R.	World Nuclear Association
Arvela, H.	Radiation and Nuclear Safety Authority (STUK), Finland
Awatsuji, Y.	Ministry of Education, Culture, Sports, Science & Technology (MEXT), Japan
Båckström, T.	Swedish Radiation Protection Authority (SSI), Sweden
Badulin, V.	Ministry of Health, Bulgaria
Baeklandt, L.	Federal Agency for Nuclear Control (FANC), Belgium
Basurto Cázares, J.	Comisión Nacional de Seguridad Nuclear y Salvaguardias (CNSNS), Mexico
Berkovskyy, V.	International Atomic Energy Agency
Boal, T.	International Atomic Energy Agency
Bochichhio, F.	Istituto Superiore di Sanità (ISS), Italy
Bologna, L.	APAT, Italy
Borras, C.	Consultant, Radiological Physics and Health Services, Washington DC, United States of America
Böttger, A.	Federal Ministry for the Environment, Nature Conservation and Nuclear Safety, Germany
Bourguignon, M.	World Health Organization
Brewer, S.	Atomic Energy of Canada Limited (AECL), Canada
Buglova, E.	International Atomic Energy Agency
Bundy, K.	Canadian Nuclear Safety Commission, Canada
Burns, P.	Australian Radiation Protection and Nuclear Safety Agency (ARPANSA), Australia
Byron, D.	Food & Agricultural Organization of the United Nations (FAO)
Cabral Molina, W.	Ministerio de Industria, Energia y Mineria (MIEM), Uruguay
Calamosca, M.	Institute for Radiation Protection, Italy
Cancio, D.	Unidad de Protección Radiológica, Spain
Canoba, A.	Autoridad Regulatoria Nuclear, Argentina
Carboneras Martinez, P.	Empresa Nacional de Residuos Radioactivos (ENRESA), Spain
Carr, Z.	World Health Organization
Cernohlavek, N.	Austrian Agency for Health and Food Safety, Austria
Chambers, D.	SENES Consultants Limited, Canada
Charette, M.	International Source Suppliers and Producers Association (ISSPA)
Cherf, A.	International Atomic Energy Agency
Cheung, K.	International Organization for Medical Physics
Chi, C.	China Institute for Radiation Protection, China
Cho, K.	Korea Institute of Nuclear Safety (KINS), Republic of Korea
Christofides, S.	International Organization for Medical Physics
Clement, C.	International Commission on Radiological Protection
Colgan, T.	International Atomic Energy Agency
Cool, D.	Nuclear Regulatory Commission, United States of America

Cooper, J.	Health Protection Agency (HPA), United Kingdom
Coppee, G.	International Labour Organization
Crick, M.	United Nations Scientific Committee on the Effects of Atomic Radiation
Cripwell, B.	International Labour Organization
Cruz-Suarez, R.	International Atomic Energy Agency
Currivan, L.	Radiological Protection Institute of Ireland (RPII), Ireland
Czarwinski, R.	International Atomic Energy Agency
de la Fuenta Puch, A.	Centro Nacional de Seguridad Nuclear (CNSN), Cuba
Deboodt, P.	International Atomic Energy Agency
Delattre, D.	International Atomic Energy Agency
Delves, D.	International Atomic Energy Agency
Dimitriou, P.	Greek Atomic Energy Commission (GAEC), Greece
Ditto, M.	Federal Ministry for Health and Women, Austria
Ebdon-Jackson, S.	Health Protection Agency (HPA), United Kingdom
Fenton, D.	Radiological Protection Institute of Ireland (RPII), Ireland
Fischer, H.	Bundesministerium für Land und Forstwirtschaft, Umwelt und Wasserwirtschaft, Austria
Frullani, S.	Instituto Superiore di Sanità (ISS), Italy
Fujii, K.	Ministry of Education, Culture, Sports, Science & Technology (MEXT), Japan
Fundarek, P.	Canadian Nuclear Safety Commission (CNSC), Canada
Garcia-Talavera, M.	Consejo de Seguridad Nuclear, Spain
Garcier, Y.	World Nuclear Association
Gaunt, M.	International Labour Organization, International Organization of Employers
Ghovanlou, A.	Health & Physics Safety, United States of America
Gilley, N.	FL Bureau of Radiation Control, United States of America
Gomaa, M.	Atomic Energy Authority (AEA), Egypt
Gonzalez, A.	Autoridad Regulatoria Nuclear, Argentina
Griebel, J.	Bundesamt für Strahlenschutz (BfS), Germany
Groth, S.	World Health Organization
Gruson, M.	Office fédéral de la santé publique, Switzerland
Guven, M.	Turkish Atomic Energy Authority, Turkey
Hamani, W.	Centre National des Sciences et Technologies Nucléaires (CNSTN), Tunisia
Hammer, J.	Swiss Federal Nuclear Safety Inspectorate (HSK), Switzerland
Hanninen, R.	Radiation and Nuclear Safety Authority (STUK), Finland
Hattori, T.	Central Research Institute of Electric Power Industry, Japan
Havukainen, R.	Radiation & Nuclear Safety Authority (STUK), Finland
Hedemann, P.	Danish Decommissioning, Denmark
Helming, M.	Federal Ministry for the Environment, Nature Conservation and Nuclear Safety, Germany
Hesse, J.	World Nuclear Association
Hoffmann, B.	Bundesamt für Strahlenschutz (BfS), Germany
Homma, T.	Japan Atomic Energy Agency (JAEA), Japan
Huffman, D.	AREVA Resources Canada Inc., Canada
Hugron, R.	National Defence Headquarters, Canada
Hulka, J.	SÚRO National Radiation Protection Institute, Czech Republic
Hunt, J.	International Atomic Energy Agency
Iimoto, T.	University of Tokyo, Japan
Inokuchi, T.	Nuclear Safety Commission, Japan
Ishikawa, N.	Nuclear Safety Commission (NSC), Japan
Ito, K.	Japan Atomic Energy Agency (JAEA), Japan

Janssens, A.	European Commission, Luxembourg
Janzekovic, H.	Slovenian Nuclear Safety Administration, Slovenia
Jensen, L.	National Institute of Radiation Protection, Denmark
Jerachanchai, S.	Office of Atoms for Peace, Thailand
Jimenez, P.	Pan American Health Organization (PAHO)
Jones, G.	International Atomic Energy Agency
Jung, T.	Bundesamt für Strahlenschutz (BfS), Germany
Jurina, V.	Public Health Authority, Slovakia
Kamenopoulou, V.	Greek Atomic Energy Commission (GAEC), Greece
Kardan, M.	Atomic Energy Organization of Iran (AEOI), Iran
Kelly, N.	United Kingdom
Kenigsberg, J	National Commission of Radiation Protection, Belarus
Kirchner, G.	Bundesamt für Strahlenschutz (BfS), Germany
Koblinger, L.	Hungarian Atomic Energy Authority (HAEA), Hungary
Koc, J.	Temelin Nuclear Power Plant, Czech Republic
Koch, J.	Soreq Nuclear Research Center, Israel
Kolovou, M.	Greek Atomic Energy Commission (GAEC), Greece
Kralik, I.	State Office for Radiation Protection, Croatia
Krca, S.	State Office for Radiation Protection, Croatia
Kuhlen, J.	Federal Ministry for the Environment, Nature Conservation and Nuclear Safety, Germany
Kulich, V.	Dukovany Nuclear Power Plant, Czech Republic
Kutkov, V.	Russian Research Centre "Kurchatov Institute", Russian Federation
Landfermann, H.	Federal Ministry for the Environment, Nature Conservation and Nuclear Safety, Germany
Larsson, C.	Swedish Radiation Protection Authority (SSI), Sweden
Lazo, E.	OECD Nuclear Energy Agency (NEA), France
Le Guen, B.	World Nuclear Association
Le Heron, J.	International Atomic Energy Agency
Lecomte, J-F.	Institut de radioprotection et de sûreté nucléaire (IRSN), France
Lindvall, C.	Barseback Kraft AB, Sweden
Lipsztein, J	Instituto de Radioproteção e Dosimetria (IRD/CNEN), Brazil
Long, K.	Food & Agricultural Organization of the United Nations (FAO)
Long, W.	Center for Radon and Air Toxics, United States of America
Lopes Gonzalez, F.	Universidad Nacional Autonoma de Nicaragua, Nicaragua
Lorenz, B.	World Nuclear Association
Louvat, D.	International Atomic Energy Agency
Lund, I.	Swedish Radiation Protection Authority (SSI), Sweden
Magnusson, S.	Icelandic Radiation Protection Institute, Iceland
Makarovska, O.	State Nuclear Regulatory Committee, Ukraine
Mansoux, H.	International Atomic Energy Agency
Marechal, N.	Commissão Nacional de Energia Nuclear (CNEN), Brazil
Marengo, M.	World Federation of Nuclear Medicine and Biology
Maringer, F.	Low-level Counting Laboratory Arsenal, Austria
Markkanen, M.	Radiation and Nuclear Safety Authority (STUK), Finland
Martin Calvarro, J.	Consejo de Seguridad Nuclear (CSN), Spain
Martincic, R.	International Atomic Energy Agency
Mason, C.	International Atomic Energy Agency
Massera, G.	Autoridad Regulatoria Nuclear, Argentina
Mayya, Y.	Bhabha Atomic Research Centre, India
McClelland, V.	Department of Energy, United States of America
McKenna, T.	International Atomic Energy Agency
McLaughlin, J.	University College Dublin, Ireland

Meghzifene, A.	International Atomic Energy Agency
Merta, A.	National Atomic Energy Agency (PAA), Poland
Metcalf, P.	International Atomic Energy Agency
Mirsaidov, U.	Nuclear and Radiation Safety Agency, Tajikistan
Miyazaki, S.	Kansai Electric Power Co. Inc., Japan
Mizumachi, W.	Japan Nuclear Energy Safety Organization (JNES), Japan
Mokrani, Z.	Nuclear Research Centre, Algeria
Mrabit, K.	International Atomic Energy Agency
Mundigl, S.	European Commission, Luxembourg
Naegele, J.	European Commission, Luxembourg
Nandakumar, A.	International Atomic Energy Agency
Niu, S.	International Labour Organization
Owen, D.	International Labour Organization
Parkes, R.	Health & Safety Executive (HSE), United Kingdom
Pather, T	National Nuclear Regulator, South Africa
Paynter, R.	Health Protection Agency (HPA), United Kingdom
Peñalosa, A.	International Organization of Employers (IOE)
Perez, M.	World Health Organization
Perrin, M.	Autorite de Sûreté Nucléaire (ASN), France
Petrova, K.	State Office for Nuclear Safety, Czech Republic
Philpott, L.	Health & Safety Executive (HSE), United Kingdom
Pinak, M.	OECD Nuclear Energy Agency (NEA), France
Poffijn, A.	Agence Federale de Controle Nucleaire (FANC), Belgium
Purvis, C.	Canadian Nuclear Safety Commission (CNSC), Canada
Radolic, V.	University Josip Juraj Strossmayer, Croatia
Rannou, A.	Institut de radioprotection et de sûreté nucléaire (IRSN), France
Rehani, M.	International Atomic Energy Agency
Ringertz, H.	International Society of Radiology
Rochedo, E.	Instituto de Radioproteção e Dosimetria (IRD/CNEN), Brazil
Rotaru, I.	National Commission for Nuclear Activities Control, Romania
Rudjord, A.	Norwegian Radiation Protection Authority, Norway
Runova, J.	Scientific and Engineering Centre for Nuclear and Radiation Safety (SEC NRS), Russian Federation
Ryder, G.	International Trade Union Confederation, Belgium
Saint-Pierre, S.	World Nuclear Association, United Kingdom
Sallit, G.	Department for Transport (DFT), United Kingdom
Salomon, S.	Australian Radiation Protection and Nuclear Safety Agency (ARPANSA), Australia
Sanz Alduan, M.	Consejo de Seguridad Nuclear (CSN), Spain
Schmitt-Hannig, A.	Bundesamt für Strahlenschutz (BfS), Germany
Sefzig, R.	Federal Ministry for the Environment, Nature Conservation and Nuclear Safety, Germany
Shannoun, F.	World Health Organization
Simeonov, G.	European Commission, Luxembourg
Sinaga, M.	Nuclear Energy Regulatory Agency (BAPETEN), Indonesia
Stasiunaitiene, R.	Ministry of Health, Lithuania
Stephen, P.	Nuclear Directorate, UK
Stern, W.	International Atomic Energy Agency
Storrie, R.	International Source Suppliers and Producers Association (ISSPA)
Sugier, A.	Institut de radioprotection et de sûreté nucléaire (IRSN), France
Suman, H.	International Atomic Energy Agency
Sutej, T.	Ministry of Health, Romania
Svensson, H.	International Organization for Medical Physics
Syahrir	National Nuclear Energy Agency (BATAN), Indonesia

Telleria, D.	International Atomic Energy Agency
Thomas, G.	Health & Safety Executive (HSE), United Kingdom
Tirmarche, M.	Institut de radioprotection et de sûreté nucléaire (IRSN), France
Todorov, N.	Bulgarian Nuclear Regulatory Agency (BNRA), Bulgaria
Tokonami, S.	National Institute of Radiological Sciences, Japan
Tomasek, L.	SÚRO National Radiation Protection Institute, Czech Republic
Tonhauser, W.	International Atomic Energy Agency
Ugleveit, F.	Norwegian Radiation Protection Authority (NRPA), Norway
Valentin, J.	International Commission on Radiological Protection
Van der Steen, J.	Nuclear Research & Consultancy Group, Netherlands
Viktorsson, C.	International Atomic Energy Agency
Wambersie, A.	International Commission on Radiation Units, Belgium
Wangler, M.	International Atomic Energy Agency
Weiss, W.	Bundesamt für Strahlenschutz (BfS), Germany
Wheatley, J.	International Atomic Energy Agency
Wiklund, A.	European Commission, Luxembourg
Wirth, E.	Bundesamt für Strahlenschutz (BfS), Germany
Wood, P.	International Society of Radiographers and Radiological Technologists
Wrixon, A.	International Atomic Energy Agency
Wymer, D.	International Atomic Energy Agency
Xiao, X.	China Institute of Atomic Energy (CIAE), China
Yonehara, H.	National Institute of Radiological Sciences (NIRS), Japan
Zafmanjato, J.	Ministère de l'éducation national et de la recherché scientifique (MENRS), Madagascar
Zeeb, H.	World Health Organization
Zodiates, T.	International Labour Organization, International Trade Union Confederation
Zuur, C.	Ministry of Housing, Spatial Planning and Environment (VROM), Netherlands

SCHEDULE III

TABLES III-1 AND III-2

These tables are included in the PDF file in the accompanying CD-ROM attached to the inside back cover of the printed publication.

1. TABLE III-1A. CONVERSION COEFFICIENTS FROM AIR KERMA FREE-IN-AIR TO $H_p(10,0^o)$ IN AN ICRU SLAB (PHOTONS)

2. TABLE III-1B. CONVERSION COEFFICIENTS FROM AIR KERMA FREE-IN-AIR TO $H_p(0.07,0^o)$ IN AN ICRU SLAB (PHOTONS)

3. TABLE III-1C. EFFECTIVE DOSE PER UNIT NEUTRON FLUENCE E/Φ for MONOENERGETIC NEUTRONS INCIDENT IN ISO GEOMETRY ON AN ADULT ANTHROPOMORPHIC COMPUTATIONAL PHANTOM

4. TABLE III-1D. REFERENCE CONVERSION COEFFICIENTS FROM FLUENCE TO DIRECTIONAL DOSE EQUIVALENT FOR MONOENERGETIC ELECTRONS AND NORMAL INCIDENCE

5. TABLE III-2A. WORKERS: COMMITTED EFFECTIVE DOSE PER UNIT INTAKE e(g) VIA INHALATION AND INGESTION ($Sv.Bq^{-1}$)

6. TABLE III-2B. COMPOUNDS AND VALUES OF GUT TRANSFER FACTOR f_l USED TO CALCULATE COMMITTED EFFECTIVE DOSE PER UNIT INTAKE VIA INGESTION FOR WORKERS

7. TABLE III-2C. COMPOUNDS, LUNG ABSORPTION TYPES AND VALUES OF GUT TRANSFER FACTOR f_l USED TO CALCULATE COMMITTED EFFECTIVE DOSE PER UNIT INTAKE VIA INHALATION FOR WORKERS

8. TABLE III-2D. MEMBERS OF THE PUBLIC: COMMITTED EFFECTIVE DOSE PER UNIT INTAKE e(g) VIA INGESTION ($Sv.Bq^{-1}$)

9. TABLE III-2E. MEMBERS OF THE PUBLIC: COMMITTED EFFECTIVE DOSE PER UNIT INTAKE e(g) VIA INHALATION ($Sv.Bq^{-1}$)

10. TABLE III-2F. LUNG ABSORPTION TYPES USED TO CALCULATE COMMITTED EFFECTIVE DOSE PER UNIT INTAKE VIA INHALATION FOR EXPOSURE TO PARTICULAR AEROSOLS OR TO GASES AND VAPOURS FOR MEMBERS OF THE PUBLIC

11. TABLE III-2G. INHALATION: COMMITTED EFFECTIVE DOSE PER UNIT INTAKE e(g) FOR SOLUBLE OR REACTIVE GASES AND VAPOURS

12. TABLE III-2H. EFFECTIVE DOSE RATE FOR EXPOSURE TO INERT GASES FOR ADULTS

BODIES FOR THE ENDORSEMENT OF IAEA SAFETY STANDARDS

An asterisk denotes a corresponding member. Corresponding members receive drafts for comment and other documentation but they do not generally participate in meetings. Two asterisks denote an alternate.

Commission on Safety Standards

Argentina: González, A.J.; *Australia*: Loy, J.; *Belgium*: Samain, J.-P.; *Brazil*: Vinhas, L.A.; *Canada*: Jammal, R.; *China*: Liu Hua; *Egypt*: Barakat, M.; *Finland*: Laaksonen, J.; *France*: Lacoste, A.-C. (Chairperson); *Germany*: Majer, D.; *India*: Sharma, S.K.; *Israel*: Levanon, I.; *Japan*: Fukushima, A.; *Korea, Republic of*: Choul-Ho Yun; *Lithuania*: Maksimovas, G.; *Pakistan*: Rahman, M.S.; *Russian Federation*: Adamchik, S.; *South Africa*: Magugumela, M.T.; *Spain*: Barceló Vernet, J.; *Sweden*: Larsson, C.M.; *Ukraine*: Mykolaichuk, O.; *United Kingdom*: Weightman, M.; *United States of America*: Virgilio, M.; *Vietnam*: Le-chi Dung; *IAEA*: Delattre, D. (Coordinator); *Advisory Group on Nuclear Security*: Hashmi, J.A.; *European Commission*: Faross, P.; *International Nuclear Safety Group*: Meserve, R.; *International Commission on Radiological Protection*: Holm, L.-E.; *OECD Nuclear Energy Agency*: Yoshimura, U.; *Safety Standards Committee Chairpersons*: Brach, E.W. (TRANSSC); Magnusson, S. (RASSC); Pather, T. (WASSC); Vaughan, G.J. (NUSSC).

Nuclear Safety Standards Committee

Algeria: Merrouche, D.; *Argentina*: Waldman, R.; *Australia*: Le Cann, G.; *Austria*: Sholly, S.; *Belgium*: De Boeck, B.; *Brazil*: Gromann, A.; **Bulgaria*: Gledachev, Y.; *Canada*: Rzentkowski, G.; *China*: Jingxi Li; *Croatia*: Valčić, I.; **Cyprus*: Demetriades, P.; *Czech Republic*: Šváb, M.; *Egypt*: Ibrahim, M.; *Finland*: Järvinen, M.-L.; *France*: Feron, F.; *Germany*: Wassilew, C.; *Ghana*: Emi-Reynolds, G.; **Greece*: Camarinopoulos, L.; *Hungary*: Adorján, F.; *India*: Vaze, K.; *Indonesia*: Antariksawan, A.; *Iran, Islamic Republic of*: Asgharizadeh, F.; *Israel*: Hirshfeld, H.; *Italy*: Bava, G.; *Japan*: Kanda, T.; *Korea, Republic of*: Hyun-Koon Kim; *Libyan Arab Jamahiriya*: Abuzid, O.; *Lithuania*: Demčenko, M.; *Malaysia*: Azlina Mohammed Jais; *Mexico*: Carrera, A.; *Morocco*: Soufi, I.; *Netherlands*: van der Wiel, L.; *Pakistan*: Habib, M.A.; *Poland*: Jurkowski, M.; *Romania*: Biro, L.; *Russian Federation*: Baranaev, Y.; *Slovakia*: Uhrik, P.; *Slovenia*: Vojnovič, D.; *South Africa*: Leotwane, W.; *Spain*: Zarzuela, J.; *Sweden*: Hallman, A.; *Switzerland*: Flury, P.; *Tunisia*: Baccouche, S.; *Turkey*: Bezdegumeli, U.; *Ukraine*: Shumkova, N.; *United Kingdom*: Vaughan, G.J. (Chairperson); *United States of America*: Mayfield, M.; *Uruguay*: Nader, A.; *European Commission*: Vigne, S.; *FORATOM*: Fourest, B.; *IAEA*: Feige, G. (Coordinator); *International Electrotechnical Commission*: Bouard, J.-P.; *International Organization for Standardization*: Sevestre, B.; *OECD Nuclear Energy Agency*: Reig, J.; **World Nuclear Association*: Borysova, I.

Radiation Safety Standards Committee

**Algeria*: Chelbani, S.; *Argentina*: Massera, G.; *Australia*: Melbourne, A.; **Austria*: Karg, V.; *Belgium*: van Bladel, L.; *Brazil*: Rodriguez Rochedo, E.R.; **Bulgaria*: Katzarska, L.; *Canada*: Clement, C.; *China*: Huating Yang; *Croatia*: Kralik, I.; **Cuba*: Betancourt Hernandez, L.; **Cyprus*: Demetriades, P.; *Czech Republic*: Petrova, K.; *Denmark*: Øhlenschlæger, M.; *Egypt*: Hassib, G.M.; *Estonia*: Lust, M.; *Finland*: Markkanen, M.; *France*: Godet, J. L.; *Germany*: Helming, M.; *Ghana*: Amoako, J.; **Greece*: Kamenopoulou, V.; *Hungary*: Koblinger, L.; *Iceland*: Magnusson, S. (Chairperson); *India*: Sharma, D.N.; *Indonesia*: Widodo, S.; *Iran, Islamic Republic of*: Kardan, M.R.; *Ireland*: Colgan, T.; *Israel*: Koch, J.; *Italy*: Bologna, L.; *Japan*: Kiryu, Y.; *Korea, Republic of*: Byung-Soo Lee; **Latvia*: Salmins, A.; *Libyan Arab Jamahiriya*: Busitta, M.; *Lithuania*: Mastauskas, A.; *Malaysia*: Hamrah, M.A.; *Mexico*: Delgado Guardado, J.; *Morocco*: Tazi, S.; *Netherlands*: Zuur, C.; *Norway*: Saxebol, G.; *Pakistan*: Ali, M.; *Paraguay*: Romero de Gonzalez, V.; *Philippines*: Valdezco, E.; *Poland*: Merta, A.; *Portugal*: Dias de Oliveira, A.M.; *Romania*: Rodna, A.; *Russian Federation*: Savkin, M.; *Slovakia*: Jurina, V.; *Slovenia*: Sutej, T.; *South Africa*: Olivier, J.H.I.; *Spain*: Amor Calvo, I.; *Sweden*: Almen, A.; *Switzerland*: Piller, G.;

**Thailand*: Suntarapai, P.; *Tunisia*: Chékir, Z.; *Turkey*: Okyar, H.B.; *Ukraine*: Pavlenko, T.; *United Kingdom*: Robinson, I.; *United States of America*: Lewis, R.; **Uruguay*: Nader, A.; *European Commission*: Janssens, A.; *Food and Agriculture Organization of the United Nations*: Byron, D.; *IAEA*: Boal, T. (Coordinator); *International Commission on Radiological Protection*: Valentin, J.; *International Electrotechnical Commission*: Thompson, I.; *International Labour Office*: Niu, S.; *International Organization for Standardization*: Rannou, A.; *International Source Suppliers and Producers Association*: Fasten, W.; *OECD Nuclear Energy Agency*: Lazo, T.E.; *Pan American Health Organization*: Jiménez, P.; *United Nations Scientific Committee on the Effects of Atomic Radiation*: Crick, M.; *World Health Organization*: Carr, Z.; *World Nuclear Association*: Saint-Pierre, S.

Transport Safety Standards Committee

Argentina: López Vietri, J.; **Capadona, N.M.; *Australia*: Sarkar, S.; *Austria*: Kirchnawy, F.; *Belgium*: Cottens, E.; *Brazil*: Xavier, A.M.; *Bulgaria*: Bakalova, A.; *Canada*: Régimbald, A.; *China*: Xiaoqing Li; *Croatia*: Belamarić, N.; **Cuba*: Quevedo Garcia, J.R.; **Cyprus*: Demetriades, P.; *Czech Republic*: Ducháček, V.; *Denmark*: Breddam, K.; *Egypt*: El-Shinawy, R.M.K.; *Finland*: Lahkola, A.; *France*: Landier, D.; *Germany*: Rein, H.; *Nitsche, F.; **Alter, U.; *Ghana*: Emi-Reynolds, G.; **Greece*: Vogiatzi, S.; *Hungary*: Sáfár, J.; *India*: Agarwal, S.P.; *Indonesia*: Wisnubroto, D.; *Iran, Islamic Republic of*: Eshraghi, A.; *Emamjomeh, A.; *Ireland*: Duffy, J.; *Israel*: Koch, J.; *Italy*: Trivelloni, S.; **Orsini, A.; *Japan*: Hanaki, I.; *Korea, Republic of*: Dae-Hyung Cho; *Libyan Arab Jamahiriya*: Kekli, A.T.; *Lithuania*: Statkus, V.; *Malaysia*: Sobari, M.P.M.; **Husain, Z.A.; *Mexico*: Bautista Arteaga, D.M.; **Delgado Guardado, J.L.; **Morocco*: Allach, A.; *Netherlands*: Ter Morshuizen, M.; **New Zealand*: Ardouin, C.; *Norway*: Hornkjøl, S.; *Pakistan*: Rashid, M.; **Paraguay*: More Torres, L.E.; *Poland*: Dziubiak, T.; *Portugal*: Buxo da Trindade, R.; *Russian Federation*: Buchelnikov, A.E.; *South Africa*: Hinrichsen, P.; *Spain*: Zamora Martin, F.; *Sweden*: Häggblom, E.; **Svahn, B.; *Switzerland*: Krietsch, T.; *Thailand*: Jerachanchai, S.; *Turkey*: Ertürk, K.; *Ukraine*: Lopatin, S.; *United Kingdom*: Sallit, G.; *United States of America*: Boyle, R.W.; Brach, E.W. (Chairperson); *Uruguay*: Nader, A.; *Cabral, W.; *European Commission*: Binet, J.; *IAEA*: Stewart, J.T. (Coordinator); *International Air Transport Association*: Brennan, D.; *International Civil Aviation Organization*: Rooney, K.; *International Federation of Air Line Pilots' Associations*: Tisdall, A.; **Gessl, M.; *International Maritime Organization*: Rahim, I.; *International Organization for Standardization*: Malesys, P.; *International Source Supplies and Producers Association*: Miller, J.J.; **Roughan, K.; *United Nations Economic Commission for Europe*: Kervella, O.; *Universal Postal Union*: Bowers, D.G.; *World Nuclear Association*: Gorlin, S.; *World Nuclear Transport Institute*: Green, L.

Waste Safety Standards Committee

Algeria: Abdenacer, G.; *Argentina*: Biaggio, A.; *Australia*: Williams, G.; **Austria*: Fischer, H.; *Belgium*: Blommaert, W.; *Brazil*: Tostes, M.; **Bulgaria*: Simeonov, G.; *Canada*: Howard, D.; *China*: Zhimin Qu; *Croatia*: Trifunovic, D.; *Cuba*: Fernandez, A.; *Cyprus*: Demetriades, P.; *Czech Republic*: Lietava, P.; *Denmark*: Nielsen, C.; *Egypt*: Mohamed, Y.; *Estonia*: Lust, M.; *Finland*: Hutri, K.; *France*: Rieu, J.; *Germany*: Götz, C.; *Ghana*: Faanu, A.; *Greece*: Tzika, F.; *Hungary*: Czoch, I.; *India*: Rana, D.; *Indonesia*: Wisnubroto, D.; *Iran, Islamic Republic of*: Assadi, M.; *Zarghami, R.; *Iraq*: Abbas, H.; *Israel*: Dody, A.; *Italy*: Dionisi, M.; *Japan*: Matsuo, H.; *Korea, Republic of*: Won-Jae Park; **Latvia*: Salmins, A.; *Libyan Arab Jamahiriya*: Elfawares, A.; *Lithuania*: Paulikas, V.; *Malaysia*: Sudin, M.; *Mexico*: Aguirre Gómez, J.; **Morocco*: Barkouch, R.; *Netherlands*: van der Shaaf, M.; *Pakistan*: Mannan, A.; **Paraguay*: Idoyaga Navarro, M.; *Poland*: Wlodarski, J.; *Portugal*: Flausino de Paiva, M.; *Slovakia*: Homola, J.; *Slovenia*: Mele, I.; *South Africa*: Pather, T. (Chairperson); *Spain*: Sanz Aludan, M.; *Sweden*: Frise, L.; *Switzerland*: Wanner, H.; **Thailand*: Supaokit, P.; *Tunisia*: Bousselmi, M.; *Turkey*: Özdemir, T.; *Ukraine*: Makarovska, O.; *United Kingdom*: Chandler, S.; *United States of America*: Camper, L.; **Uruguay*: Nader, A.; *European Commission*: Necheva, C.; *European Nuclear Installations Safety Standards*: Lorenz, B.; **European Nuclear Installations Safety Standards*: Zaiss, W.; *IAEA*: Siraky, G. (Coordinator); *International Organization for Standardization*: Hutson, G.; *International Source Suppliers and Producers Association*: Fasten, W.; *OECD Nuclear Energy Agency*: Riotte, H.; *World Nuclear Association*: Saint-Pierre, S.

No. 22

IAEA
International Atomic Energy Agency

Where to order IAEA publications

In the following countries IAEA publications may be purchased from the sources listed below, or from major local booksellers. Payment may be made in local currency or with UNESCO coupons.

AUSTRALIA
DA Information Services, 648 Whitehorse Road, MITCHAM 3132
Telephone: +61 3 9210 7777 • Fax: +61 3 9210 7788
Email: service@dadirect.com.au • Web site: http://www.dadirect.com.au

BELGIUM
Jean de Lannoy, avenue du Roi 202, B-1190 Brussels
Telephone: +32 2 538 43 08 • Fax: +32 2 538 08 41
Email: jean.de.lannoy@infoboard.be • Web site: http://www.jean-de-lannoy.be

CANADA
Bernan Associates, 4501 Forbes Blvd, Suite 200, Lanham, MD 20706-4346, USA
Telephone: 1-800-865-3457 • Fax: 1-800-865-3450
Email: customercare@bernan.com • Web site: http://www.bernan.com

Renouf Publishing Company Ltd., 1-5369 Canotek Rd., Ottawa, Ontario, K1J 9J3
Telephone: +613 745 2665 • Fax: +613 745 7660
Email: order.dept@renoufbooks.com • Web site: http://www.renoufbooks.com

CHINA
IAEA Publications in Chinese: China Nuclear Energy Industry Corporation, Translation Section, P.O. Box 2103, Beijing

CZECH REPUBLIC
Suweco CZ, S.R.O., Klecakova 347, 180 21 Praha 9
Telephone: +420 26603 5364 • Fax: +420 28482 1646
Email: nakup@suweco.cz • Web site: http://www.suweco.cz

FINLAND
Akateeminen Kirjakauppa, PO BOX 128 (Keskuskatu 1), FIN-00101 Helsinki
Telephone: +358 9 121 41 • Fax: +358 9 121 4450
Email: akatilaus@akateeminen.com • Web site: http://www.akateeminen.com

FRANCE
Form-Edit, 5, rue Janssen, P.O. Box 25, F-75921 Paris Cedex 19
Telephone: +33 1 42 01 49 49 • Fax: +33 1 42 01 90 90
Email: formedit@formedit.fr • Web site: http://www. formedit.fr

Lavoisier SAS, 145 rue de Provigny, 94236 Cachan Cedex
Telephone: + 33 1 47 40 67 02 • Fax +33 1 47 40 67 02
Email: romuald.verrier@lavoisier.fr • Web site: http://www.lavoisier.fr

GERMANY
UNO-Verlag, Vertriebs- und Verlags GmbH, Am Hofgarten 10, D-53113 Bonn
Telephone: + 49 228 94 90 20 • Fax: +49 228 94 90 20 or +49 228 94 90 222
Email: bestellung@uno-verlag.de • Web site: http://www.uno-verlag.de

HUNGARY
Librotrade Ltd., Book Import, P.O. Box 126, H-1656 Budapest
Telephone: +36 1 257 7777 • Fax: +36 1 257 7472 • Email: books@librotrade.hu

INDIA
Allied Publishers Group, 1st Floor, Dubash House, 15, J. N. Heredia Marg, Ballard Estate, Mumbai 400 001,
Telephone: +91 22 22617926/27 • Fax: +91 22 22617928
Email: alliedpl@vsnl.com • Web site: http://www.alliedpublishers.com

Bookwell, 2/72, Nirankari Colony, Delhi 110009
Telephone: +91 11 23268786, +91 11 23257264 • Fax: +91 11 23281315
Email: bookwell@vsnl.net

ITALY
Libreria Scientifica Dott. Lucio di Biasio "AEIOU", Via Coronelli 6, I-20146 Milan
Telephone: +39 02 48 95 45 52 or 48 95 45 62 • Fax: +39 02 48 95 45 48
Email: info@libreriaaeiou.eu • Website: www.libreriaaeiou.eu

JAPAN
Maruzen Company, Ltd., 13-6 Nihonbashi, 3 chome, Chuo-ku, Tokyo 103-0027
Telephone: +81 3 3275 8582 • Fax: +81 3 3275 9072
Email: journal@maruzen.co.jp • Web site: http://www.maruzen.co.jp

REPUBLIC OF KOREA
KINS Inc., Information Business Dept. Samho Bldg. 2nd Floor, 275-1 Yang Jae-dong SeoCho-G, Seoul 137-130
Telephone: +02 589 1740 • Fax: +02 589 1746 • Web site: http://www.kins.re.kr

NETHERLANDS
De Lindeboom Internationale Publicaties B.V., M.A. de Ruyterstraat 20A, NL-7482 BZ Haaksbergen
Telephone: +31 (0) 53 5740004 • Fax: +31 (0) 53 5729296
Email: books@delindeboom.com • Web site: http://www.delindeboom.com

Martinus Nijhoff International, Koraalrood 50, P.O. Box 1853, 2700 CZ Zoetermeer
Telephone: +31 793 684 400 • Fax: +31 793 615 698
Email: info@nijhoff.nl • Web site: http://www.nijhoff.nl

Swets and Zeitlinger b.v., P.O. Box 830, 2160 SZ Lisse
Telephone: +31 252 435 111 • Fax: +31 252 415 888
Email: infoho@swets.nl • Web site: http://www.swets.nl

NEW ZEALAND
DA Information Services, 648 Whitehorse Road, MITCHAM 3132, Australia
Telephone: +61 3 9210 7777 • Fax: +61 3 9210 7788
Email: service@dadirect.com.au • Web site: http://www.dadirect.com.au

SLOVENIA
Cankarjeva Zalozba d.d., Kopitarjeva 2, SI-1512 Ljubljana
Telephone: +386 1 432 31 44 • Fax: +386 1 230 14 35
Email: import.books@cankarjeva-z.si • Web site: http://www.cankarjeva-z.si/uvoz

SPAIN
Díaz de Santos, S.A., c/ Juan Bravo, 3A, E-28006 Madrid
Telephone: +34 91 781 94 80 • Fax: +34 91 575 55 63
Email: compras@diazdesantos.es, carmela@diazdesantos.es, barcelona@diazdesantos.es, julio@diazdesantos.es
Web site: http://www.diazdesantos.es

UNITED KINGDOM
The Stationery Office Ltd, International Sales Agency, PO Box 29, Norwich, NR3 1 GN
Telephone (orders): +44 870 600 5552 • (enquiries): +44 207 873 8372 • Fax: +44 207 873 8203
Email (orders): book.orders@tso.co.uk • (enquiries): book.enquiries@tso.co.uk • Web site: http://www.tso.co.uk

On-line orders
DELTA Int. Book Wholesalers Ltd., 39 Alexandra Road, Addlestone, Surrey, KT15 2PQ
Email: info@profbooks.com • Web site: http://www.profbooks.com

Books on the Environment
Earthprint Ltd., P.O. Box 119, Stevenage SG1 4TP
Telephone: +44 1438748111 • Fax: +44 1438748844
Email: orders@earthprint.com • Web site: http://www.earthprint.com

UNITED NATIONS
Dept. I004, Room DC2-0853, First Avenue at 46th Street, New York, N.Y. 10017, USA
(UN) Telephone: +800 253-9646 or +212 963-8302 • Fax: +212 963-3489
Email: publications@un.org • Web site: http://www.un.org

UNITED STATES OF AMERICA
Bernan Associates, 4501 Forbes Blvd., Suite 200, Lanham, MD 20706-4346
Telephone: 1-800-865-3457 • Fax: 1-800-865-3450
Email: customercare@bernan.com · Web site: http://www.bernan.com

Renouf Publishing Company Ltd., 812 Proctor Ave., Ogdensburg, NY, 13669
Telephone: +888 551 7470 (toll-free) • Fax: +888 568 8546 (toll-free)
Email: order.dept@renoufbooks.com • Web site: http://www.renoufbooks.com

Orders and requests for information may also be addressed directly to:

Marketing and Sales Unit, International Atomic Energy Agency
Vienna International Centre, PO Box 100, 1400 Vienna, Austria
Telephone: +43 1 2600 22529 (or 22530) • Fax: +43 1 2600 29302
Email: sales.publications@iaea.org • Web site: http://www.iaea.org/books

11-39241